Genetics of Alzheimer's Disease

Genetics of Alzheimer's Disease

Editors

Justin Miller
Laura Ibanez

MDPI • Basel • Beijing • Wuhan • Barcelona • Belgrade • Manchester • Tokyo • Cluj • Tianjin

Editors

Justin Miller
Sanders-Brown Center on Aging
University of Kentucky
Lexington
United States

Laura Ibanez
Department of Psychiatry and
Department of Neurology
NeuroGenomics and
Informatics Center
Washington University in St. Louis
St. Louis
United States

Editorial Office
MDPI
St. Alban-Anlage 66
4052 Basel, Switzerland

This is a reprint of articles from the Special Issue published online in the open access journal *Genes* (ISSN 2073-4425) (available at: www.mdpi.com/journal/genes/special_issues/genetics_alzheimer).

For citation purposes, cite each article independently as indicated on the article page online and as indicated below:

LastName, A.A.; LastName, B.B.; LastName, C.C. Article Title. *Journal Name* **Year**, *Volume Number*, Page Range.

ISBN 978-3-0365-2933-2 (Hbk)
ISBN 978-3-0365-2932-5 (PDF)

© 2022 by the authors. Articles in this book are Open Access and distributed under the Creative Commons Attribution (CC BY) license, which allows users to download, copy and build upon published articles, as long as the author and publisher are properly credited, which ensures maximum dissemination and a wider impact of our publications.

The book as a whole is distributed by MDPI under the terms and conditions of the Creative Commons license CC BY-NC-ND.

Contents

About the Editors vii

Laura Ibanez and Justin B. Miller
Editorial for the Genetics of Alzheimer's Disease Special Issue: October 2021
Reprinted from: *Genes* **2021**, *12*, 1794, doi:10.3390/genes12111794 1

Laura Ibanez, Carlos Cruchaga and Maria Victoria Fernández
Advances in Genetic and Molecular Understanding of Alzheimer's Disease
Reprinted from: *Genes* **2021**, *12*, 1247, doi:10.3390/genes12081247 5

Erik D. Huckvale, Matthew W. Hodgman, Brianna B. Greenwood, Devorah O. Stucki, Katrisa M. Ward and Mark T. W. Ebbert et al.
Pairwise Correlation Analysis of the Alzheimerrsquo;s Disease Neuroimaging Initiative (ADNI) Dataset Reveals Significant Feature Correlation
Reprinted from: *Genes* **2021**, *12*, 1661, doi:10.3390/genes12111661 31

Megan Torvell, Sarah M. Carpanini, Nikoleta Daskoulidou, Robert A. J. Byrne, Rebecca Sims and B. Paul Morgan
Genetic Insights into the Impact of Complement in Alzheimerrsquo;s Disease
Reprinted from: *Genes* **2021**, *12*, 1990, doi:10.3390/genes12121990 43

Maria Garofalo, Cecilia Pandini, Daisy Sproviero, Orietta Pansarasa, Cristina Cereda and Stella Gagliardi
Advances with Long Non-Coding RNAs in Alzheimer's Disease as Peripheral Biomarker
Reprinted from: *Genes* **2021**, *12*, 1124, doi:10.3390/genes12081124 59

Benjamin C. Shaw, Yuriko Katsumata, James F. Simpson, David W. Fardo and Steven Estus
Analysis of Genetic Variants Associated with Levels of Immune Modulating Proteins for Impact on Alzheimer's Disease Risk Reveal a Potential Role for SIGLEC14
Reprinted from: *Genes* **2021**, *12*, 1008, doi:10.3390/genes12071008 67

Eun-Gyung Lee, Sunny Chen, Lesley Leong, Jessica Tulloch and Chang-En Yu
TOMM40 RNA Transcription in Alzheimer's Disease Brain and Its Implication in Mitochondrial Dysfunction
Reprinted from: *Genes* **2021**, *12*, 871, doi:10.3390/genes12060871 79

Robert J. van der Linden, Ward De Witte and Geert Poelmans
Shared Genetic Etiology between Alzheimer's Disease and Blood Levels of Specific Cytokines and Growth Factors
Reprinted from: *Genes* **2021**, *12*, 865, doi:10.3390/genes12060865 99

Sarah M. Carpanini, Janet C. Harwood, Emily Baker, Megan Torvell, The GERAD1 Consortium and Rebecca Sims et al.
The Impact of Complement Genes on the Risk of Late-Onset Alzheimer's Disease
Reprinted from: *Genes* **2021**, *12*, 443, doi:10.3390/genes12030443 109

Devanshi Patel, Xiaoling Zhang, John J. Farrell, Kathryn L. Lunetta and Lindsay A. Farrer
Set-Based Rare Variant Expression Quantitative Trait Loci in Blood and Brain from Alzheimer Disease Study Participants
Reprinted from: *Genes* **2021**, *12*, 419, doi:10.3390/genes12030419 121

About the Editors

Justin Miller

Dr. Justin Miller completed his Ph.D. and postdoctoral work at Brigham Young University, where he analyzed the evolutionary history of codon usage biases across the Tree of Life and applied bioinformatics to identify genetic variants associated with Alzheimer's disease. He is currently an Assistant Professor in Biomedical Informatics at the University of Kentucky and affiliated with the Sanders-Brown Center on Aging.

Laura Ibanez

Dr. Laura Ibanez received a Ph.D. in Biomedicine from the University of Barcelona, Spain and completed a postdoctoral fellowship at Washington University School of Medicine in St. Louis (WUSM). She joined the faculty at WUSM in 2020, and is currently an Assistant Professor of Psychiatry and Neurology at the NeuroGenomics and Informatics Center (NGI-Center). She combines state-of-the-art genomic and bioinformatics approaches to study the biology of neurodegenerative diseases. Her research interests are currently focused on developing new minimally invasive tools for rapid and accurate diagnosis of neurodegenerative diseases to provide early detection and improve management.

Editorial

Editorial for the Genetics of Alzheimer's Disease Special Issue: October 2021

Laura Ibanez [1,2,*] and Justin B. Miller [3,*]

1. Department of Psychiatry, Washington University in Saint Louis, St. Louis, MO 63110, USA
2. Department of Neurology, Washington University in Saint Louis, St. Louis, MO 63110, USA
3. Sanders-Brown Center on Aging, University of Kentucky, Lexington, KY 40536, USA
* Correspondence: ibanezl@wustl.edu (L.I.); justin.miller@uky.edu (J.B.M.)

Citation: Ibanez, L.; Miller, J.B. Editorial for the Genetics of Alzheimer's Disease Special Issue: October 2021. *Genes* **2021**, *12*, 1794. https://doi.org/10.3390/genes12111794

Received: 25 October 2021
Accepted: 29 October 2021
Published: 14 November 2021

Publisher's Note: MDPI stays neutral with regard to jurisdictional claims in published maps and institutional affiliations.

Copyright: © 2021 by the authors. Licensee MDPI, Basel, Switzerland. This article is an open access article distributed under the terms and conditions of the Creative Commons Attribution (CC BY) license (https://creativecommons.org/licenses/by/4.0/).

Alzheimer's disease is a complex and multifactorial condition regulated by both genetics and lifestyle, which ultimately results in the accumulation of β-amyloid (Aβ) and tau proteins in the brain, loss of gray matter, and neuronal death. This Special Issue, entitled "Genetics of Alzheimer's Disease," focuses on genetic contributions to this debilitating disease that may lead to better targeted therapeutics and diagnostics. This issue contains six original research articles and two review papers that further our collective knowledge of Alzheimer's disease etiology and genetic risk factors underlying the disease.

Ibanez, Cruchaga [1] present a snapshot of current work in Alzheimer's disease genetics by contextualizing recent advances targeting immune responses from *CD33* and *TREM2* with more contemporary approaches investigating the amyloid cascade hypothesis. They conclude that the molecular mechanisms modulating Alzheimer's disease pathogenesis should be considered broadly, and identifying additional genetic predisposition variants may lead to better treatment through early prediction and diagnosis. Shaw, Katsumata [2] found that limitations inherent with stringent multiple testing correction may mask the ability to detect Alzheimer's disease risk from complex copy number variations and genes with coupled expression within immunomodulatory tyrosine-phosphorylated inhibitory motifs (ITIMs) or activation motifs (ITAMs). They show that protein quantitative trait loci associate with Alzheimer's disease more frequently for genes encoding ITIM/ITAM family members than non-ITIM/ITAM genes. Additionally, mitochondrial dysfunction may play a role in Alzheimer's disease, and differential transcription of the Translocase of Outer Mitochondria Membrane 40 (*TOMM40*) gene is associated with Alzheimer's disease in postmortem brains [3].

Data mining is often used to identify novel disease-associated genetic risk factors for Alzheimer's disease. Huckvale, Hodgman [4] describe challenges with data mining on data from the Alzheimer's Disease Neuroimaging Initiative that may cause issues with various machine learning algorithms. They describe significant feature correlation, where >90% of all biomarkers are significantly correlated with at least one other biomarker in that dataset. They recommend removing highly correlated features before performing large-scale data analyses. Carpanini, Harwood [5] performed a targeted analysis of Alzheimer's disease-associated genes within the complement system in the IGAP dataset, and they confirmed genetic associations for both *CLU* and *CR1*, but *C1S* was not significantly associated with Alzheimer's disease. They conclude that larger genome-wide association datasets and long-read sequencing technologies may help better characterize the complement system genetic landscape and its role in Alzheimer's disease risk.

Peripheral blood biomarkers offer a promising, non-intrusive mechanism for early Alzheimer's disease diagnosis. Garofalo, Pandini [6] explore how non-coding RNA molecules longer than 200 nucleotides are differentially expressed in peripheral tissue of Alzheimer's disease patients and may lead to noninvasive confirmatory targets or prognostic biomarkers. Specifically, plasma BACE1-AS levels are differentially expressed in the pre-symptomatic phase, indicating long non-coding RNA molecules may be a viable pre-symptomatic diagnostic target. Patel, Zhang [7] demonstrate that rare genetic variants significantly impact gene expression and gene co-expression in Alzheimer's disease, indicating that set-based gene analyses are necessary to fully capture gene dynamics related to disease progression. Those pathway-level analyses confirmed substantial immune and inflammatory expression quantitative trait loci associated with Alzheimer's disease, as suggested in the review published in this Special Issue [1]. An association between immune markers and Alzheimer's disease is also proposed by van der Linden, De Witte [8], who drew a correlation between levels of blood cytokines and growth factors and Alzheimer's disease genetic risk factors. They found that eight immune markers (three growth factors and five cytokines) were downregulated by Alzheimer's disease genetic risk factors, while seven immune markers (five growth factors and three cytokines) were upregulated in the blood when Alzheimer's disease genetic risk factors were present.

The articles included in this Special Issue cover a range of topics and provide comprehensive insights to direct future Alzheimer's disease genetics research. The growing number of pathways, genes, proteins, and molecules that appear to be involved in Alzheimer's disease is of special interest, which highlights the importance of analyzing Alzheimer's disease as a systemic disease and not just a neurological disorder. Findings in the blood may serve as not only potential biomarkers or potential drug targets, but also help decipher the pathobiology of the disease. We anticipate that this Special Issue will help researchers search for additional genetic associations that will help refine our understanding of the etiology of this complex disease.

Author Contributions: J.B.M. and L.I. both contributed to writing and editing this editorial and special issue on the genetics of Alzheimer's disease. All authors have read and agreed to the published version of the manuscript.

Funding: This work was supported by BrightFocus Foundation grant #A2020118F (PI: Miller) and #A2021033S (PI: Ibanez), the Alzheimer's Drug Discovery Foundation grant #GDAPB-201807-2015632 (PI: Ibanez), the Department of Defense grant #W81XWH2010849 (PI: Ibanez) and NIH grants #1P30AG072946-01 (University of Kentucky Alzheimer's Disease Research Center), and #K99AG062723 (PI: Ibanez).

Institutional Review Board Statement: Not applicable.

Informed Consent Statement: Not applicable.

Data Availability Statement: Not applicable.

Acknowledgments: We appreciate the donors to the BrightFocus Foundation, the University of Kentucky, Washington University in St. Louis, and the Sanders-Brown Center on Aging for their contributions to support this research.

Conflicts of Interest: The authors declare no conflict of interest.

References

1. Ibanez, L.; Cruchaga, C.; Fernández, M.V. Advances in Genetic and Molecular Understanding of Alzheimer's Disease. *Genes* **2021**, *12*, 1247. [CrossRef] [PubMed]
2. Shaw, B.C.; Yuriko Katsumata, Y.; Simpson, J.F.; Fardo, D.W.; Estus, S. Analysis of Genetic Variants Associated with Levels of Immune Modulating Proteins for Impact on Alzheimer's Disease Risk Reveal a Potential Role for SIGLEC14. *Genes* **2021**, *12*, 1008. [CrossRef] [PubMed]
3. Lee, E.-G.; Chen, S.; Leong, L.; Tulloch, J.; Yu, C.-E. TOMM40 RNA Transcription in Alzheimer's Disease Brain and Its Implication in Mitochondrial Dysfunction. *Genes* **2021**, *12*, 871. [CrossRef] [PubMed]
4. Huckvale, E.D.; Hodgman, M.W.; Greenwood, B.B.; Stucki, D.O.; Ward, K.M.; Ebbert, M.T.W.; Kauwe, J.S.K. Pairwise Correlation Analysis of the Alzheimer's Disease Neuroimaging Initiative (ADNI) Dataset Reveals Significant Feature Correlation. *Genes* **2021**, *12*, 1661. [CrossRef]
5. Carpanini, S.M.; Harwood, J.C.; Baker, E.; Torvell, M.; Sims, R.; Williams, J.; Morgan, B.P.; The GERAD1 Consortium. The Impact of Complement Genes on the Risk of Late-Onset Alzheimer's Disease. *Genes* **2021**, *12*, 443. [CrossRef] [PubMed]
6. Garofalo, M.; Pandini, C.; Sproviero, D.; Pansarasa, O.; Cereda, C.; Gagliardi, S. Advances with Long Non-Coding RNAs in Alzheimer's Disease as Peripheral Biomarker. *Genes* **2021**, *12*, 1124. [CrossRef] [PubMed]

7. Patel, D.; Zhang, X.; Farrell, J.J.; Lunetta, K.L.; Farre, L.A. Set-Based Rare Variant Expression Quantitative Trait Loci in Blood and Brain from Alzheimer Disease Study Participants. *Genes* **2021**, *12*, 419. [CrossRef] [PubMed]
8. van der Linden, R.J.; de Witte, W.; Poelmans, G. Shared Genetic Etiology between Alzheimer's Disease and Blood Levels of Specific Cytokines and Growth Factors. *Genes* **2021**, *12*, 865. [CrossRef] [PubMed]

Review

Advances in Genetic and Molecular Understanding of Alzheimer's Disease

Laura Ibanez [1,2], Carlos Cruchaga [1,2,3] and Maria Victoria Fernández [1,2,*]

1. Department of Psychiatry, Washington University School of Medicine, 660 S. Euclid Ave. B8134, St. Louis, MO 63110, USA; ibanezl@wustl.edu (L.I.); cruchagac@wustl.edu (C.C.)
2. Neurogenomics and Informatics Center, Washington University School of Medicine, 660 S. Euclid Ave. B811, St. Louis, MO 63110, USA
3. Hope Center for Neurological Disorders, Washington University School of Medicine, 660 S. Euclid Ave. B8111, St. Louis, MO 63110, USA
* Correspondence: fernandezv@wustl.edu

Abstract: Alzheimer's disease (AD) has become a common disease of the elderly for which no cure currently exists. After over 30 years of intensive research, we have gained extensive knowledge of the genetic and molecular factors involved and their interplay in disease. These findings suggest that different subgroups of AD may exist. Not only are we starting to treat autosomal dominant cases differently from sporadic cases, but we could be observing different underlying pathological mechanisms related to the amyloid cascade hypothesis, immune dysfunction, and a tau-dependent pathology. Genetic, molecular, and, more recently, multi-omic evidence support each of these scenarios, which are highly interconnected but can also point to the different subgroups of AD. The identification of the pathologic triggers and order of events in the disease processes are key to the design of treatments and therapies. Prevention and treatment of AD cannot be attempted using a single approach; different therapeutic strategies at specific disease stages may be appropriate. For successful prevention and treatment, biomarker assays must be designed so that patients can be more accurately monitored at specific points during the course of the disease and potential treatment. In addition, to advance the development of therapeutic drugs, models that better mimic the complexity of the human brain are needed; there have been several advances in this arena. Here, we review significant, recent developments in genetics, omics, and molecular studies that have contributed to the understanding of this disease. We also discuss the implications that these contributions have on medicine.

Keywords: Alzheimer Disease; amyloid β; tau; APOE; TREM2; neuroinflammation; OMICS; biomarkers; therapeutics

Citation: Ibanez, L.; Cruchaga, C.; Fernández, M.V. Advances in Genetic and Molecular Understanding of Alzheimer's Disease. *Genes* **2021**, *12*, 1247. https://doi.org/10.3390/genes12081247

Academic Editor: Italia Di Liegro

Received: 13 May 2021
Accepted: 10 August 2021
Published: 15 August 2021

Publisher's Note: MDPI stays neutral with regard to jurisdictional claims in published maps and institutional affiliations.

Copyright: © 2021 by the authors. Licensee MDPI, Basel, Switzerland. This article is an open access article distributed under the terms and conditions of the Creative Commons Attribution (CC BY) license (https://creativecommons.org/licenses/by/4.0/).

1. Introduction

Ever since Alois Alzheimer provided the first clinical and pathological description of this disease in 1901, we have learned that Alzheimer's disease (AD) is a complex and multifactorial condition in which the interplay of both genetic (65%) and lifestyle (35%) factors [1] is involved in the accumulation of protein aggregates of β-amyloid (Aβ) and tau in the brain that ultimately causes neuronal death and loss of gray matter. AD has had an estimated cost to the United States healthcare system of USD 290 billion. Disease prevalence is expected to grow from 5.8 million in 2019 to 14 million by 2050 [2]; hence, extensive international research efforts have been devoted to deciphering the causes of disease and developing therapeutics that may alter the course of the disease. Results have been elusive for several reasons.

There are three main etiological categories in AD: autosomal dominant AD (ADAD), early onset AD (EOAD), and late-onset AD (LOAD). Mutations in one of the three genes with Mendelian inheritance that cause disease (amyloid precursor protein (*APP*) and presenilin 1 and 2 (*PSEN1*, *PSEN2*)) are normally present in the ADAD form, with early onset (before 65 years old) and rapid progression. This form is fairly rare, about 1% of cases, but it has been instrumental for our initial understanding of the pathology of the disease, the development of animal models, and the design of the first therapeutic treatments. *APP*, *PSEN1*, and *PSEN2* are members of the same Aβ processing pathway. The identification of specific mutations directly related to the main pathological hallmark of AD, extracellular aggregates of Aβ plaques, led to great advances in our understanding of the disease and to the formulation of the amyloid cascade hypothesis [3]. The amyloid cascade hypothesis states that a malfunction in the system causes an accumulation of Aβ in the brain that triggers a cascade of events, ultimately resulting in cell death.

The remaining 99% of cases are largely classified into EOAD (~5%) or LOAD (~95%) according to the age of disease onset, with a threshold arbitrarily established at 65 years old. In addition, these can also be further categorized into sporadic AD (sAD) or familial AD (fAD), depending on the incidence of cases within families. Unless specified, for the remainder of the text, we will refer to the non-ADAD forms (EOAD, LOAD, sAD, and fAD) as AD. The non-ADAD forms present a more complex genetic architecture, with associations to over 29 genetic loci identified to date [4–7]. The loci identified through genetic studies have suggested alternative pathways beyond those involved in Aβ accumulation, such as tau aggregation, lipid metabolism, the innate immune response, and endosomal vesicle recycling. It is not clear whether any of these pathways have a greater role than the others. On top of this complexity, microglia are active players in the clearance of Aβ plaques whose activation seems to be regulated by APOE; yet, hyper activation of microglia is detrimental [8,9] (Figure 1).

This complexity raises questions for the "one-size-fits-all" approach. Critics of the amyloid cascade hypothesis have stated that the failure of Aβ-targeted drugs is partly due to the fact that ADAD may be different from AD. As such, a plethora of potential drug targets have been envisioned, but most have been unsuccessful for various reasons. First, it is unclear how and when the implicated genes and pathways interact and if they are "active" in all individuals. Second, a definitive diagnosis of AD cannot be made without confirmation by autopsy, so physicians and scientists have to rely on biomarkers (e.g., measuring Aβ, tau, or p-tau in cerebrospinal fluid (CSF) or plasma, or Aβ deposition in the brain using positron emission tomography (PET) imaging) to make diagnoses as accurate as possible. However, these methods are either not fully implemented (plasma), invasive (lumbar puncture for CSF), or expensive (CSF and imaging), which limits their generalized use in screening of trial participants. These diagnostic challenges lead to a "contamination" of clinical trials with non-AD cases, mostly with misdiagnosed frontotemporal dementia (FTD) cases [10] and clinically diagnosed cases that were amyloid-negative by Pittsburgh compound B (PIB) imaging or CSF ELISA (around 30% are amyloid-negative) [11]. Another major problem is that pathological changes that underlie brain degeneration and cognitive loss begin at least 10 to 20 years before dementia onset [12,13]. Most clinical trials so far have focused on individuals with clinical symptoms, in which the neurodegeneration may be too advanced for any therapeutic to reverse or stop deterioration [14]. Accordingly, current clinical trials are trying to include mild cognitive impairment (MCI) cases, defined as a transitional state between normal aging and dementia, although not all MCI patients convert to AD. Hence, a critical goal of biomedical research is to identify biomarkers of AD for these preclinical stages allowing for early diagnosis and intervention.

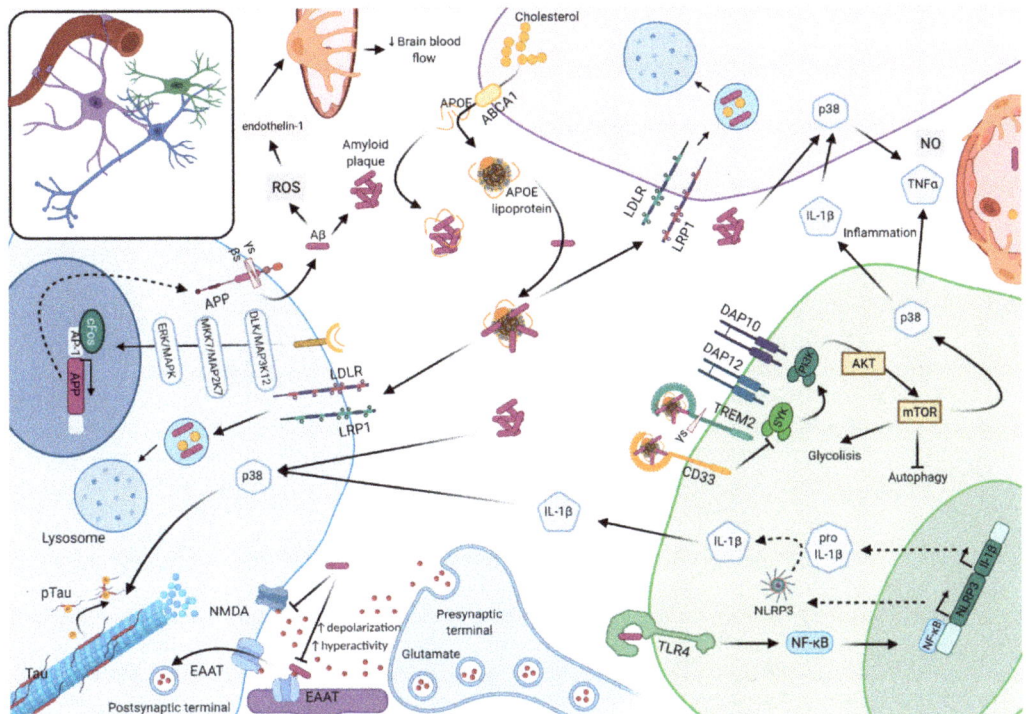

Figure 1. Schematic representation of the molecular interplay between neurons, astrocytes, microglia, and vasculature system in Alzheimer's disease. Neurons produce the transmembrane amyloid precursor protein (APP). APP is cleaved by γ (γs) and β secretase (βs) into amyloid β (Aβ) units of different length aggregates extracellularly into plaques. Oligomeric Aβ promotes the generation of ROS that triggers the release of endothelin-1, causing perycite constriction, which decreases brain blood flow. Soluble Aβ blocks the reuptake of synaptically released glutamate by either N-methyl-D-aspartate receptor (NMDA) or by excitatory amino acid transporter (EAAT) receptors causing glutamate accumulation perisynaptically (excitotoxicity), which increases depolarization and promotes hyperactivity. In microglia, Aβ binds to the toll-like receptor 4 (TLR4), which causes translocation of the nuclear factor kappa-light-chain-enhancer of activated B cells (NF-κB) from the cytosol to the nucleus, where it increases the transcription of NLRP3 and pro-IL-1β. In the cytoplasm, via activated caspase-1, the inflammasome promotes the maturation of IL-1β. Amyloid plaques stimulate the activation of p38MAPK (p38) in microglia, astrocyte, and neuron. In microglia, p38 activation results in upregulation of proinflammatory cytokines, interleukin 1β (IL-1β) and tumor necrosis factor α (TNFα); IL-1β in turns activates p38 in astrocytes and neurons. In astrocytes, p38 activation causes increased expression of TNFα and nitric oxide (NO) and excitotoxicity. In neurons, p38 activation results in tau phosphorylation (Tau → p-Tau) and microtubule disassembly. APOE is mostly generated by astrocytes; free APOE can facilitate Aβ blood bran barrier (BBB) transit, but it can also accelerate aggregation and deposition of Aβ in an isoform-dependent manner. APOE can be lipidated by ABCA1 transporter-forming lipoprotein particles that bind soluble Aβ, which are then uptaken by neurons and glia via cell-surface receptors, including low-density lipoprotein receptor (LDLR) and low-density lipoprotein receptor-related protein (LRP1), and degraded at the lysosome. When free APOE binds to ApoE receptors in neurons, it can activate a non-canonical MAPK pathway, in an isoform-dependent manner, that induces cFOS phosphorylation stimulating the transcription factor AP-1, which in turn enhances transcription of APP. The complex amyloid plaque lipidated APOE can stimulate microglia through transmembrane proteins triggering receptor expressed on myeloid cells 2 (TREM2) and sialic acid-binding Ig-like lectin 3 (CD33). TREM2 activation induces the association of TREM2 to DAP12, which gets phosphorylated and recruits spleen tyrosine kinase (SYK), which activates phosphoinositide 3-kinase (PI3K) that depends on DAP10. PI3K targets protein kinase B (AKT) and activates the mammalian target of rapamycin (mTOR), which activates glycolysis, the p38MAPK pathway, and inhibits autophagy. Instead, CD33 activation inhibits PI3K. The complement receptor 1 (CR1) is a receptor for the complement components C3b and C4b and promotes the phagocytosis of Aβ.

2. There Is More to Alzheimer's Disease Than Amyloid

2.1. The Amyloid vs. Tau Hypotheses

The identification of mutations in the *APP*, *PSEN1*, and *PSEN2* genes in families with ADAD led to the formulation of the amyloid cascade hypothesis. The presenilin genes encode secretases (α, β, and γ) that cleave APP, a transmembrane protein, into amyloid β (Aβ) units of different lengths (from 36 to 43 amino acids in length) that are released to the extracellular space. Neurons are the main producers of Aβ, and mutations in these genes cause an overproduction of Aβ_{42} and its various toxic forms that accumulate into plaques. Plaque formation may start a series of events involving synaptic dysfunction by interfering with glutamatergic synapsis and inflammation by causing microglia hyperactivity, which promotes hyperphosphorylation of tau. Tau hyperphosphorylation can lead to the generation of destabilized microtubules in the intracellular space that aggregate and form neurofibrillary tangles (NFTs), leading to widespread neuronal dysfunction and death [15]. AD is a disease that starts with the accumulation of Aβ plaques followed by the formation of NFTs, which would be more likely to cause the observed neuronal dysfunction and degeneration [16,17] since the spreading of tau pathology is highly correlated with the patterns of clinical symptoms and cognitive decline [18]. Nonetheless, a decrease in cerebral blood flow is one of the first changes in AD pathology and could reflect dysfunction of contractile pericytes. Nortley et al. (2019) measured capillary diameters at positions near pericytes in human brain biopsies from cognitively unimpaired individuals with Aβ plaques, as well as in AD mice models (APP^{NL-G-F}). They observed that capillaries were constricted near pericytes and that this constriction was correlated with the severity of Aβ deposition. In addition, oligomeric Aβ promotes the generation of reactive oxygen species (ROS) (NOX4), which triggers the release of endothelin-1, which acts on ET$_A$ receptors to induce pericyte contraction. However, it is not clear what damage to synapses and neurons is due to the decrease in energy supply caused by Aβ-induced capillary constriction [19].

Another early feature of AD caused by Aβ depositions is neuronal hyperactivity. Zoo et al. (2019) demonstrated that, given a neuron-specific baseline activity driven by glutamatergic synapses, soluble Aβ blocks the reuptake of synaptically released glutamate, causing presynaptic glutamate accumulation, which increases depolarization and promotes hyperactivity [20]. Current AD treatment with memantine blocks the effects of excess glutamate that inhibits signal detection by NMDA glutamate receptors. This study suggests that targeting excitatory amino acid transporters (EAAT) may be a mechanism to therapeutically target neuronal hyperactivation at the early stages of the disease.

Immune response and inflammation are other key features in the pathology of AD (as later discussed in Section 2.3, "The underground of Alzheimer's disease"). Upon microglia activation by Aβ deposits, the NLRP3 inflammasome assembles and initiates an inflammatory response, which contributes to the seeding and spreading of Aβ in AD mouse models [21]. Ising et al. (2019) demonstrated that in the absence of the NLRP3 inflammasome, tau hyperphosphorylation and aggregation were reduced, suggesting that tau pathology is a downstream process of the Aβ cascade and dependent on microglia activation [22].

Critics of the amyloid cascade hypothesis suggest that amyloid could be a side-effect of the disease [15] and that AD could be a disorder that is triggered by impairment of APP metabolism but progresses through tau-related pathology rather than Aβ-related pathology [16]. He et al. (2018) injected tau derived from human AD brains (AD-tau) into AD transgenic mice, overexpressing pathogenic Aβ. Mice showed accumulation of AD-tau seeds within dystrophic axons surrounding Aβ plaques; these seeds spread to neuronal somas and dendrites to recruit endogenous soluble tau and form NFTs and neuropil threads (NTs) [23]. Tau phosphorylation is mediated by neuronal p38 mitogen-activated protein kinase (p38MAPK), which is activated by Aβ plaques and cytokines (IL-1β). There are four isoforms of p38MAPK (α, β, δ, and γ), and each can phosphorylate tau at specific sites. Maphis et al. (2016) found that selective suppression of the p38αMAPK rescued late-stage tau pathology and improved working memory in 20 months old mice expressing human tau (hTau) [24]. On the other hand, Ittner et al. (2016) observed that depletion of p38γMAPK in APP23 mice increased cognitive deficits whereas increased expression of p38γMAPK (i.e., increased tau phosphorylation) abolished those deficits. In addition, they observed that APP23.p38γ$^{-/-}$.tau$^{-/-}$ mice did not present memory deficits, suggesting that the effects of p38γ were tau-dependent [25]. While no mutations have been found within the MAPK pathway that are associated with AD, somatic mutations in the *BRAF* gene (which is part of the MAPK pathway) in the erythro-myeloid progenitor lineage in mice may cause neurodegeneration [26].

Finally, Klein et al. (2019) studied the histone 3 lysine 9 acetylation (H3K9ac) mark in 669 aged brains from the Religious Order Study (ROS) and the Rush Memory and Aging Project (MAP) and correlated it to their Aβ and tau pathological signatures. Almost 23% of H3K9ac domains were associated with tau protein load, whereas only 2% were associated with Aβ. The tau-associated domains clustered in large genomic regions within gene promoter or enhancer regions and in open chromatin compartments. Using induced pluripotent stem cell (iPSC)-derived neurons, they further showed that overexpression of *MAPT*, without tangle formation, is enough to induce chromatin reorganization, suggesting that the tau effects in epigenomic architecture are an early event in tau pathology [27].

2.2. Polyvalent APOE

ApoE is a protein that transports lipids from one tissue or cell to another. It is highly expressed in the liver, adipose tissue, and artery wall, but it is also found in the central nervous system (CNS), where it is mainly synthesized by astrocytes and microglia [28]. Two SNPs (rs429358 and rs7412) within the *APOE* gene generate three major allelic variants (ε2, ε3, and ε4), which have a worldwide frequency of 8.4%, 77.9%, and 13.7%, respectively [29]. These isoforms bind to lipids, receptors, and Aβ with varying efficiencies [28,30–32]. The presence of the APOE ε4 allele has been associated with hyperlipidemia and hypercholesterolemia [33,34]; one copy of ε4 increases risk for AD by ~3-fold and two copies by ~12-fold [35], yet only 40% of sporadic AD and 60–70% of LOAD families carry this allele [29]. In addition, having the ε4 allele correlates with an average of 2–5 years earlier AAO, or up to 10 years if carrying two copies of the ε4 allele [36,37]. This risk not only applies to sporadic or familial LOAD but also to ADAD [38]. The ε2 allele is considered protective and would delay the appearance of symptoms [36,39,40]. The ε3 allele has a neutral effect, although rare mutations associated with this isoform (*APOE3*-Christchurch p.R136S) confer protection against developing the disease when occurring in homozygosis [41].

It has been suggested that APOE contributes to AD pathology through both Aβ-dependent and Aβ-independent pathways. In an isoform-dependent manner, free APOE can influence Aβ deposition, but it can also help soluble Aβ to cross the blood-brain barrier (BBB) [42–44]. Alternatively, lipidated APOE recruits soluble Aβ preventing Aβ plaque formation, but also facilitates its cell-absorption by neprilysin, produced by microglia, or by cell-surface receptors (LRP1, LDLR, and HSPG) where it is degraded at the lysosomes [45–47]. However, recent studies suggest that APOE secreted by glia stimulates *APP* transcription and Aβ production in neurons in an isoform-dependent manner [48].

On the other hand, *APOE* has been associated with CSF tau levels [49–51]. iPSC-derived neurons expressing ApoE ε4, but not ApoE ε3, had higher levels of tau phosphorylation [52]. Similarly, tau transgenic mice that express human *APOE* had higher tau levels in the brain and a greater extent of somatodendritic tau redistribution compared to *Apoe*$^{-/-}$ mice [52]. More importantly, through gene editing, Wang et al. [53] converted *ApoE* ε4 to *ApoE* ε3 and was able to rescue the normal phenotype.

Finally, beyond its effect on amyloid and tau, ApoE also influences microglial activation, the latter possibly through Trem2 interaction [54,55]. According to Krasemann et al. [55], the transition of microglia from a homeostatic- to a disease-associated microglial (DAM) phenotype would be dependent on ApoE. Supporting this, Ulrich et al. (2018) found that Apoe deficient mice presented a significant reduction in fibrillar plaque-associated microgliosis and activated microglial gene expression [8]. Recently, Parhizkar et al. (2019)

showed that amyloid plaque seeding is increased in the absence of functional Trem2 and that this seeding goes along with decreased microglial clustering and reduced plaque-associated ApoE [56]. Yet, it is uncertain how Trem2 interferes with microglial lipid metabolism [57].

2.3. The Underground of Alzheimer's Disease—The Immune System

Early genome-wide association studies (GWAs) were successful at identifying additional genetic risk factors for AD, such as *CLU*, *PICALM*, *CR1*, *BIN1*, and *CD33* [4,58–60]. The immune pathway was seen as an important component of AD pathology since *CLU*, *CR1*, and *CD33* have putative functions in the immune system. More recent studies with larger data sets identified additional genome-wide significant genes involved in the immune pathway, including *MS4A*, *CD2AP*, *EPHA1*, and *ABCA7* [61,62]. The later discovery of loss-of-function variants in *TREM2* provided scientists with particular targets to focus on in the study of the immune response in AD pathology [63,64]. More recently, it was found that the minor allele of rs1057233 (G), near the GWAS *CELF1* risk locus [4], showed an association with lower expression of *SPI1* in monocytes and macrophages [65]. *SPI1* encodes PU.1, a microglial transcription factor critical for myeloid cell development, which regulates the expression of numerous AD risk genes (*TREM2*, *TYROBP*, *CD33*, *MS4A* cluster genes, and *ABCA7*) [54,65]. Recent genome-wide meta-analyses of AD-by-proxy individuals identified 29 risk loci that are strongly expressed in immune-related tissues and cell types [6]. Two of these genes, *ADAM10* and *ACE*, along with *TREM2* and *SPI1*, were found to have a genome-wide significant association in the largest known GWAS that included around 95,000 people [7]. *ADAM10* is the α-secretase for APP that produces a secreted ectodomain fragment (sAPPα) that has neuroprotective and neurotrophic properties. In addition, ADAM10 cleaves Notch and various immune and growth factor proteins [66]. *ACE* encodes an enzyme involved in the conversion of angiotensin I into a physiologically active peptide, angiotensin II, a potent vasopressor. ACE is also involved in Aβ degradation [67]. It is still unclear how mutations in these genes relate to microglial dysfunction, but overexpression of *ACE* in microglia and macrophages in a double transgenic mice model for AD (APPswe/PS1dE9) substantially reduced cerebral soluble A$β_{42}$, vascular and parenchymal Aβ deposits, and astrocytosis [68].

Activation of p38MAPK signaling in microglia (due to Aβ plaques) releases proinflammatory cytokines in astrocytes and neurons, resulting in inflammation and tau phosphorylation [69]. Deficits in *TREM2* have also been linked to dysregulation in PPARγ/p38MAPK signaling. Microglia switch from using oxidative phosphorylation for energy production to glycolysis in the presence of Aβ plaques. This metabolic reprogramming depends on the mTOR-HIF-1α pathway [70]. Piers et al. (2019) observed that iPSC-derived microglia from patients carrying pathogenic *TREM2* mutations had trouble switching to glycolytic metabolism, which ultimately was reflected by dysregulation of the PPARγ/p38MAPK signaling [71].

Microglia are also stimulated by Aβ plaques through transmembrane proteins CD33 and TREM2. While CD33 activation dampens microglial phagocytosis by inhibiting phosphatidylinositol-3 kinase (PI3K), TREM2 responds to ligand binding by activating PI3K to increase phagocytosis [72]. Functional analysis suggests that downregulation of *CD33* may be beneficial to AD since amyloid levels were reduced in a mouse model of AD (APPswe/PS1dE9) that were also $CD33^{-/-}$ [73,74]. However, the consequence of regulating *TREM2* expression is unclear. For example, higher soluble TREM2 (sTREM2) in MCI or AD individuals was associated with reduced rates of cognitive decline and clinical progression [75]. *Trem2* knockout in a mouse model of tauopathy (PS19) resulted in a reduction in neurodegeneration and inflammation [76]. However, loss of Trem2 function increased the seeding and spread of neuritic plaque aggregates in mouse models of AD (APPPS1-21) injected with human AD-tau [77]. In contrast, overexpression of *TREM2* in BV-2 cells (an immortalized murine microglial cell line) promoted clearance of Aβ products and mediated neuroinflammation by downregulating the expression of inflammatory factors [78]. These apparently conflicting roles for *TREM2*, protective vs. harmful, could be due to the disease stage examined in each study [79].

Studies in 5XFAD mouse models indicate that *TREM2* is essential for microglia to acquire a DAM phenotype. However, in human AD, the DAM signature of microglia seems to be conditioned by the expression of the *IRF8* transcription factor. Loss-of-function mutations in *TREM2* promoted a less reactive phenotype of microglia [80], suggesting that the risk-effect that *TREM2* exerts on AD may be regulated by third parties. In fact, recent studies suggest that *TREM2* could be regulated indirectly through *MS4A4A*. A common variant in the MS4A cluster (rs1582763) is associated with increased CSF sTREM2. This study also demonstrated that TREM2 is implicated in disease in general and not only in those individuals that carry *TREM2* risk variants. Mendelian randomization analyses demonstrated that high sTREM2 levels were protective. In addition, it was found that MS4A4A and TREM2 co-localize intracellularly, suggesting MS4A4A as a potential therapeutic target for AD [81]; Alector, Inc. is currently testing an antibody that mimics the protective effect of the MS4A4A variant.

Finally, it seems that microglia and the autophagy pathway may interact in the pathology of AD disease. Hung et al. (2018) described deficits in the lysosome and autophagosome pathways using iPSC-derived neurons from individuals carrying pathogenic mutations in *PSEN1* and *APP* [82]. However, the disruption of these pathways seems to be more pronounced in the LOAD forms, for which several genes have been associated [83] and in relation to the expression decline of some proteins in the autophagy pathway due to age, which is exacerbated in AD [84]. Since microglia are the main cells phagocytizing Aβ plaques, Heckmann et al. (2019) hypothesized that defects in the autophagy pathway could influence microglial behavior in AD [84]. Using the 5XFAD AD mouse model, they identified that LC3 associated with endosomal membranes (LC3-associated endocytosis—LANDO) supports the clearance of Aβ deposits and prevents microglia activation. However, this process is dependent on the presence of several autophagy regulators, including ATG5, whose expression decreases with age [84].

Another gene implicated in lysosome and autophagosome dysfunction and risk for AD is *TMEM106B*. This gene has been reported to be associated with FTD in granulin (*GRN*) mutation carriers [85] and with AD interacting with *APOE*. More recently, Li et al. used a digital deconvolution [86] to estimate the brain cell-type proportion from multiple cohorts. Genetic scans of neuronal proportion indicate that a variant located in the *TMEM106B* gene is the major regulator of neuronal proportion in adults but not young individuals [87]. Impaired lysosomal function reduces lysosomal degradative efficiency, which leads to an abnormal build-up of toxic components in the cell. An impaired lysosomal system has been associated with normal aging and a broad range of neurodegenerative disorders, including AD [87]. These findings suggest that *TMEM106B* could be a potential target for neuronal protection therapies to ameliorate cognitive and functional deficits.

3. Unraveling the Molecular Mechanisms in AD Pathogenesis

Most of our knowledge of AD genetic risk factors originated from studying blood samples; yet, the genome is transcribed and translated differentially across tissues in response to different transcription factors, metabolic signals, and environmental responses. Accordingly, in recent years there has been an effort to study different omic layers (genome, transcriptome, proteome, metabolome, and epigenome) in different tissues affected by AD (blood, plasma, CSF, and brain) and different cell types (macrophages, neurons, microglia, astrocytes, and oligodendrocytes), whether it is in human samples, mouse models of AD, or iPSCs. These studies have advanced our understanding of the roles of amyloid, tau, APOE, and the immune system in identifying the pathological triggers and order of events in pathogenesis. Recent studies suggest that DNA is not static during an individual's lifetime and is a feature of the aging brain. This DNA instability is worse in AD [88–90]. As such, the existence of somatic genetic mosaicism was suggested after detecting increased

APP copy number variants (CNV) in cortical neuronal nuclei of sporadic AD patients [91], although this event would only contribute to a small percentage of sporadic AD cases [92].

Most of our current understanding of processes downstream of the genome comes from the analysis of the RNA (in its multiple species), whether it comes from blood, bulk brain tissue, or, more recently, from specific cell types. Using whole transcriptome profiling of AD brains, over 2000 genes were found to be deregulated in AD cases [93], and most of these were associated with functional pathways involved in the immune response, apoptosis, cell proliferation, energy metabolism, and synaptic transmission, corroborating findings from previous GWAs analyses [94,95]. Yet, this transcriptome profile may differ depending on the main AD risk factor. Network analysis of transcriptomic data from AD patients identified aging-associated processes (inflammation, oxidative stress, and metabolic pathways) were differentially altered depending on *APOE* genotype (44 vs. 33). Integration of these results with GWAs data indicated an epistatic interaction between *APOE* and several genes in the Notch pathway, suggesting a possible link between *APOE* and its role in inflammation and oxidation [96].

Integration of transcriptomic data with other phenotypes can also reveal important aspects of the disease. Transcriptome-wide network analysis with longitudinal cognitive data was used to identify a set of co-expressed genes that are related to both Aβ and cognitive decline and are separate from those that cause AD pathology [97]. Using PET imaging and brain transcriptomic data, Sepulcre et al. (2018) found an association between gene expression profiles and Aβ and tau pathology progression across the cerebral cortex. Aβ propagation was related to a dendrite-related genetic profile mostly driven by the *CLU* gene; tau propagation was related to an axon-related genetic profile led by the *MAPT* gene. This study helps to clarify the possible relationships between Aβ and tau pathology. For example, *BACE1*, the gene that codifies for the β-secretase enzyme that cleaves APP, was identified as one of the central genes in the tau-related interactome network. In addition, a lipid metabolism category was identified as commonly involved in the propagation of both Aβ and tau. APOE had a dominant role; participants who were *APOE* ε4^+ had a linear relationship between the propagation pattern for Aβ and tau compared to those who were *APOE* ε4^- [98]. This suggests that a person's genetic profile may define whether the spread of pathology is due to Aβ or tau.

Bioinformatic deconvolution approaches can untangle the transcriptomic signature of bulk brain tissue and infer the relative contribution of different cell types to a particular cell expression pattern [63]. These methods revealed that carriers of pathogenic mutations in *APP*, *PSEN1*, *PSEN2*, or *APOE* presented with lower neuron and higher astrocyte proportions compared to patients with sporadic AD, suggesting that the presence of AD genetic risk factors affects the cellular composition of AD brains [86]. Technological advancements have enabled the sequencing of individual cell nuclei, which allows for the identification of cell-specific patterns. Pioneering studies using this technology in human AD brains were capable of identifying cell-type-specific transcriptomic profiles [99]. This technology also allows the mapping of specific cell profiles at certain points in time. It was found that early in the pathology, the disease-associated transcriptional changes were highly cell-type-specific, whereas, in later disease stages, the transcriptional signature of the disease was common across cell types, mostly centered around global stress response [100]. Similarly, these technologies reveal that human microglia have an AD-related gene signature that is distinct from that described in mouse models [101], suggesting that mouse models of AD may not be adequate in vivo systems to study all pathological aspects of the disease, as discussed later in this section.

Circular RNA (circRNA) are formed by back-splicing (head-to-tail splicing) of messenger RNAs during normal processing. They were first described in eukaryotic cells, and later studies suggested that they were enriched in the synapse, acting as sponges of micro RNA (miRNA). One of the events implicated in the pathophysiology of AD is synapse loss [102]. CircRNAs were also found to be co-expressed with known causal AD genes, such as *APP* and *PSEN1*, suggesting that some circRNA are also part of the causal AD pathway. CircRNA brain expression explained more about AD clinical manifestations than the number of *APOE* ε4 alleles, suggesting that cirRNA could be used as biomarkers for AD.

Epigenetics can lead to changes that affect gene activity and expression but do not require changes of the nucleotide sequence. The main epigenetic marks are DNA methylation and histone modifications. Pioneering epigenome-wide association studies (EWAs) of AD examined the hypermethylation of CpG sites in the brain cortex of AD patients [103,104]. These studies identified methylation changes in the *ANK1* gene. Cell-specific EWAs of neuron and glia single nuclei validated and assigned methylation of *ANK1* as specific to glia [105]. More recently, Smith et al. (2019) performed a targeted methylation analysis finding that differential *ANK1* methylation is a common feature across the entorhinal brain cortex of subjects with AD, Huntington's disease (HD), or Parkinson's disease (PD), but not those with vascular dementia (VD) or dementia with Lewy bodies (DLB) unless these individuals had co-existing AD pathology [106]. Other studies have looked at histone acetylation marks, H4K16ac particularly, which is an epigenetic modification of the DNA that serves to regulate chromatin compaction, gene expression, stress responses, and DNA damage repair. H4K16ac marks are usually enriched with aging, but the exploration of brain temporal lobe tissue from AD patients revealed losses of acetylated histone H4K16ac, which was superior in the proximity of genes linked to aging and with previously identified AD genetic loci [107].

Metabolic decline is one of the earliest symptoms detected in patients with MCI [108]. Hence, by identifying those metabolites that differ between MCI to AD patients, it is possible to establish panels of time-specific metabolic biomarkers, which will help us understand the mechanisms of disease at different stages. Several metabolites, such as alanine, aspartate, and glutamate, have been associated with AD and cognitive decline, whereas unsaturated fatty acids have been associated with early memory impairment [109–111]. As such, the Alzheimer's Disease Metabolomics Consortium observed that preclinical AD cases were enriched in sphingomyelins and ether-containing phosphatidylcholines compared to symptomatic AD cases in which acylcarnitines and several amines were the most representative metabolomic groups [112]. Similarly, sphingolipids were found to be the more distinct species between AD cases and controls and were associated with the severity of AD pathology at autopsy and AD progression [113,114]. The correlation between these metabolic signatures in the brain and peripheral tissues, as well as their relationship with key AD pathological biomarkers, remains to be elucidated, but they are a promising source of novel biomarkers.

Despite the potential of these novel technologies, a major challenge in advancing this line of research is access to sufficient brain tissue from different stages of the disease to explore pathology at different time points. iPSCs have become useful models to study single-cell behavior in disease [115]. Similarly, monocyte-derived microglia-like (MDMi) cells recapitulate key aspects of microglia phenotype and function, and their expression of neurodegenerative disease-related genes is different from that of monocytes [116]. These models have been useful for studying the effects of specific variants on cell phenotype, e.g., the study of ADAD mutations [117] and the effects of tau-related mutations in AD [118]. Most importantly, these studies help differentiate functional responses observed in mice from those in human systems [119]. Transcriptomic analysis of 5XFAD mice and human AD single nuclei brain cells revealed discordances in the transcriptomic signature of oligodendrocytes, astrocytes, and microglia between these two systems [80].

Yet, the brain is a complex organ involving the interaction of multiple cell types, with different proportions in different brain areas. In addition, AD pathogenesis is a combination of Aβ accumulation, phosphorylated tau (p-tau) formation, hyperactivation of glial cells, and neuronal loss. Therefore, iPSC or MDMi alone cannot be expected to model the brain response to AD pathogenic events. A novel engineered model has been developed that grows three-dimensionally interacting neurons, astrocytes, and microglia in order to model AD pathogenesis [120]. This new 3D human AD triculture system mirrored the first pathogenic AD stages, Aβ aggregation and p-tau formation, and the induction of microglia recruitment that leads to marked neuron and astrocyte loss [120]. More recently, several groups have managed to incorporate the microvasculature into these organoids, providing them with blood-brain barrier characteristics [121]. These models have the potential to advance our understanding of AD in multiple ways. First, we can study the pathological processes that occur in the brains of AD patients. By combining patient-specific iPSC with triculture 3D technology, we could evaluate the differential activation of pathways in a patient-specific manner. Subsequently, different drugs could be tested to evaluate the efficacy and occurrence of side effects in a patient-specific and personalized manner.

4. Early Prediction and Diagnosis Are Key to Better Treatment

Current diagnostic tools for AD patients include neuropsychological tests to assess memory and other cognitive abilities, whether it is Clinical Dementia Rating (CDR) [122], Mini-Mental State Exam (MMSE) [123], Montreal Cognitive Assessment (MoCA) [124], or Consortium to Establish a Registry for Alzheimer's Disease (CERAD) [125] and measurement of biomarkers in the brain, CSF, or blood. Biomarkers can be measured in the brain using imaging techniques (MRI, CT, or PET) that inform us about metabolic changes in the brain (glucose) or deposit of certain protein aggregates (Aβ, tau, and p-tau) or in biofluids (blood and CSF). A definitive diagnosis of AD can only be made by pathological exam of the brain postmortem. However, it is known that pathologic changes in the brain occur years before a person starts showing signs of cognitive impairment [126]. A key factor in the success of clinical trials and in the treatment of AD patients is the ability to intervene before clinical symptoms appear. Thus, we are in urgent need of tools that can identify individuals at risk and be applied at the population level in a fast and affordable manner. The challenge remains in having access to well-characterized, large cohorts with a longitudinal repository of body fluids in which sets of biomarkers can be investigated in a retrospective manner. In addition, as we have reviewed in this manuscript, not only Aβ and tau contribute to the pathogenesis of AD; there are many other pathways involved that have the potential to provide accurate biomarkers to predict and follow disease progression and response to potential treatments. Clinical trials for drug targets have mainly focused on reducing the production of Aβ or trying to clear its deposits from the brain, mainly based on the knowledge derived from ADAD patients. Yet, the pathological mechanisms starting and driving ADAD, EOAD, and LOAD might differ, thus, different therapeutic approaches that take into account the etiology and genetic background of each individual should be investigated. In the next section we summarize the most recent papers on risk prediction, detection, diagnosis, and treatment strategies (Table 1).

Table 1. Summary of selected recent studies of risk prediction, early diagnosis, and treatment.

Approaches to Risk Prediction				
	Ref.	Approach	Findings	
	[127]	PRS	EOAD, sLOAD, and fLOAD have different PRS profiles	
	[128]	PRS + biomarker	Prediction of conversion or AAO	
	[129]	PRS + brain atrophy + MMSE score	Better progression prediction	
	[130]	PRS + brain atrophy + MMSE score + CSF data	Individuals with high PRS and with amyloid and tau pathology showed a faster rate of memory decline, even among APOE ε4 non-carriers	
	[131]	PRS	PRS differentiate AD, FTD, PD, and ALS	
Biomarkers for Early Diagnosis				
Ref.	Target	Localization	Indicative Of	Aβ-Independent	Tau-Independent
[132]	sTREM2	CSF, plasma	Onset and progression of tau pathology	Y	N
[133]	Nfl	plasma, CSF	Cytoskeleton protein released with neuron death	Y	Y
[134]	sPDGRβ	blood	Blood-brain barrier breakdown	Y	Y

Table 1. Cont.

Possible Treatments and Clinical Trials						
Clinical Trial Ref.	Target	Mechanism	Participants	Goal	Drug	Status
NCT01677572	Amyloid β	Monoclonal antibodies	Mild AD and MCI	Clearance of Aβ plaques	Aducanumab	Approved to treat Alzheimer's disease
NCT02760602			Prodromal AD		Solanezumab	T—no evidence that prodromal AD benefits from drug
NCT02008357			Older Individuals at risk (APOE4+)		Solanezumab	P3—not recruiting
NCT01760005			DIAN-TU		Solanezumab	P3; H—no change in cognitive performance
NCT01760005			DIAN-TU		Gantereumab	P3; H—no change in cognitive performance
NCT01998841			Colombian family		Crenezumab	P3—not recruiting
NCT01661673	Y-secretase	y-modulator	Mild Cognitive Impairment	Decrease production of toxic Aβ	EVP-0962	P2—completed

PRS—polygenic risk score; Nfl—Neurofilament Light; T—terminated; H—halted; P2—phase II; P3—phase III.

4.1. Risk Prediction and Prevention

Polygenic risk scores (PRS) aim to generate a genetic profile of an individual and to predict this individual's chances of developing a certain condition. The first PRS for AD included all 21 variants identified by the IGAP consortia with a prediction accuracy of 78.2% [135], although its accuracy could reach 82% depending on the AD subtype [136]. In fact, PRS calculated for stratified AD etiologies revealed an accuracy of 75% for fAD, 72% for sAD, and 67% for AD [127]. Under the hypothesis that a pathway-specific PRS could be more powerful at predicting certain pathological aspects of the disease, Darst et al. (2016) clustered the SNPs for the AD PRS into the major AD pathways (Aβ clearance, cholesterol metabolism, and immune response) and tested their association with cognition function and AD-biomarkers (Aβ imaging, CSF Aβ, tau, and p-tau) [128]. Unfortunately, these prediction values were no more accurate than models including all known disease variants, suggesting there is room for improvement of these predictors. Kauppi et al. (2018) generated a PRS that predicted progression from MCI to AD over 120 months; when these data were combined with baseline brain atrophy score and/or MMSE score, the prediction model was significantly improved (AUC = 84%) compared to the use of the PRS alone [129]. Furthermore, adding biomarker data from CSF and imaging measurements, the same group found that individuals with high PRS and with amyloid and tau pathology showed a faster rate of cognitive decline, even among *APOE ε4* non-carriers [130]. In a similar study, it was found that adding imaging information of Aβ and tau deposition (PET) and neurodegeneration (MRI) to a model that already included clinical and genetic information improved the prediction accuracy of memory decline [137]. Despite the increase in predictive ability, it is uncertain whether these improvements will be clinically relevant for the daily practice of predicting people at risk. We are still in need of predictors that do not rely on biomarkers of already occurring pathology.

4.2. Early Detection, Diagnosis, and Prognosis

In the search for biomarkers that can reflect what is occurring in the brain, CSF has been the fluid of preference given its direct contact with the CNS. Aβ peptides and tau were among the first proteins to be investigated in CSF for detection and diagnostic purposes. This was followed by the investigation of VILIP-1 (marker of neuronal injury), YKL40 (marker of inflammation), neurogranin (NGRN—marker of synaptic function), and CLU (an apolipoprotein involved in several Aβ processes and a risk factor for AD). However, these markers are not disease-specific (VILIP-1, YKL40 [138], NGRN [139]) or do not present differently between cases and controls (e.g., CLU [140]). Therefore, there is still a need to identify biomarkers that (i) are AD-specific and can predict the onset of cognitive decline and (ii) are independent of Aβ and tau metabolism so that disease progression can be monitored in patients enrolled in clinical trials using drugs targeting Aβ or tau. Recently, progress has been made in the analysis of Aβ in blood (plasma) in order to replace screening in CSF and reduce its invasiveness and related expenses [75].

Soluble TREM2 (sTREM2) is detectable in CSF and serum. Its levels are elevated in MCI-AD compared to AD or controls and correlate with those of tau and p-tau, but not Aβ [132,141]. CSF sTREM2 levels increase before the onset of symptoms, but after amyloidosis and neuronal injury have already begun [132,142,143], suggesting that *TREM2* may play a critical role in the onset and progression of tau pathology and microglia activation. In addition, a higher ratio of CSF sTREM2 to CSF p-tau181 concentrations predicted slower conversion from cognitively normal to symptomatic stages or from MCI to AD dementia [75]. sTREM2 can be generated by the proteolytic action of ADAM10 (an α-secretase also involved in the cleavage of APP), but missense mutations in the immunoglobulin-like domain and stalk region have been found to interfere with the cleavage site and shedding of sTREM in opposite directions [144]. In addition, there are three alternative transcripts for *TREM2*. One lacks the transmembrane domain and encodes only the sTREM2 form. Using bulk brain transcriptomic data from AD cases, TREM2 carriers, and controls. del-Aguila et al. (2019) showed that up to 25% of sTREM2 may be translated from *TREM2* isoforms that lack the transmembrane domain; in addition, the expression of this particular isoform was significantly different in cases compared to controls [145]. The role of sTREM2 in the cascade of pathologic events remains unclear, and because of the lack of selective inflammatory markers, it is uncertain whether inflammation and microglial activation or tau-related abnormalities occur first. Yet again, it may be that the order of pathologic events may differ between ADAD cases and AD patients that do not carry mutations on those genes [146].

Neurofilament light chain (NfL) is an intrinsic protein of the axonal cytoskeleton that is released when neurons die. NfL was found in high concentrations in CSF and blood among participants of the Dominantly Inherited Alzheimer's Network (DIAN) ~6.8 years before the onset of symptoms [133]. NfL is not an AD-exclusive biomarker, but since it is Aβ- and tau-independent, it has the potential to be used as a proximity marker and as a marker to monitor therapy response.

The contribution of neurovascular dysfunction and blood-brain barrier (BBB) breakdown to cognitive impairment is widely recognized. These both develop early in AD; however, the relationship between vascular pathology and Aβ and tau is still unknown. Nation et al. (2019) studied the CSF from cognitively normal individuals as well as from individuals with early cognitive dysfunction who were CSF Aβ+, Aβ−, p-tau+, or p-tau−. The soluble platelet-derived growth factor receptor-β (sPDGFRβ) is mainly expressed by brain vascular mural cells but not by other cells of the CNS. Nation et al. (2019) found that sPDGFRβ was increased in the CSF of individuals with advanced CDR regardless of CSF Aβ of p-tau status, suggesting that biomarkers focused on the integrity of the brain vasculature could be a novel source for biomarkers of cognitive dysfunction in both individuals with and without AD [134].

Although most blood- and CSF-based biomarkers focus on protein levels, cell-free nucleotides are also being investigated for use in diagnostic tests. Disease-specific cell-free RNA transcripts have been found at increased levels in the blood of affected individuals [147,148]. Additionally, small non-coding miRNA were used to differentiate cases from controls across different neurodegenerative diseases: AD, FTL, and ALS were differentiated from each other with accuracy ranging from 0.77 to 0.93 [149]. These studies employed samples from symptomatic patients only, so further studies are required in preclinical individuals to confirm the potential of this miRNA-based approach as a diagnostic tool.

4.3. Treatment

The majority of ongoing phase III clinical trials were developed under the umbrella of the amyloid hypothesis and so are mainly focused on stopping the production of Aβ or aimed at clearing Aβ plaques. A set of drugs have been designed to target the γ-secretase complex with the aim to prevent the production of Aβ altogether. However, γ-secretase cleaves not only APP but also up to another 50 transmembrane protein substrates, including Notch receptors [150]. Recent studies have revealed that ADAD mutations destabilize the intermediate enzyme-substrate complexes between APP and γ-secretase, promoting early disassociation of γ-secretase from Aβ and thereby releasing longer and more amyloidogenic Aβ peptides [151]. The second wave of current clinical trials use monoclonal antibodies to promote Aβ clearance, have been designed for a very specific subset of the population, and try to tackle the disease before symptoms appear. The Dominantly Inherited Alzheimer's Network Trials Unit (DIAN-TU) selects participants from families with autosomal dominant mutations in either *APP*, *PSEN1*, or *PSEN2* genes and treats them with either Solanezumab, a monoclonal antibody that targets soluble Aβ, or Gantenerumab, a monoclonal antibody that interacts with Aβ plaques and activates microglia phagocytosis [152]. Similarly, the Alzheimer's Prevention Initiative (API) ADAD trial is focused on a large Colombian family with ADAD due to a pathogenic mutation in *PSEN1* (p.E280A); members of this family are treated with Crenezumab, a monoclonal antibody that recognizes multiple Aβ forms and stimulates amyloid phagocytosis while limiting inflammation [153]. The API also has two more trials: (i) CAD106, a vaccine that combines multiple Aβ forms and aims to produce a strong antibody response while avoiding inflammatory T cell activation, and (ii) Umibecestat, which seeks reduction Aβ production by inhibiting the BACE1 protease. These trials are conducted with 60–75-year-old cognitively normal *APOE* ε4 homozygotes and aim to prevent the appearance of disease [154]. However, the predictions for this trial are not very promising, since Verubecestat, another drug aimed at inhibiting BACE1 to block Aβ production, failed to improve the cognitive abilities of prodromal AD cases [155]. That is not the case for aducanumab (Aduhelm), a monoclonal antibody against aggregated forms of Aβ approved by the FDA in June 2021 (https://www.fda.gov/drugs/news-events-human-drugs/fdas-decision-approve-new-treatment-alzheimers-disease (accessed on 3 August 2021)). Even though it has proved effective in reducing the burden of Aβ plaques, it is still not clear if it also reduces the symptomatology [156].

Despite some anti-Aβ therapeutic drugs look promising in phase III clinical trials, recent data suggest that amyloid would be aside-effect of the brain's response to stress in sporadic AD, not a causative factor as in familial AD [157]. Cognitive decline and pathogenic events are directly associated with the initiation of tau aggregation, hence an interest in developing tau-related therapies [158]. Tau pathology in AD is characterized by a disruption of 3R to 4R tau isoforms, resulting in an approximately 2:1 4R:3R ratio [159,160]. Tau expression could be reduced with small interfering RNA (siRNA) [161] or antisense oligonucleotides (ASO) [162]. These mechanisms have not yet been tested in clinical trials for AD or other tauopathies, but they have been used for cancer [163] and spinal muscular atrophy [164].

APOE polymorphisms have been recognized to contribute to AD pathology by both gain-of- and loss-of-function properties. This bi-directional effect must be taken into account when designing therapies targeting ApoE [165]. On the one hand, mechanisms that enhance ApoE quantity have been shown to promote Aβ clearance and synaptic function in an isoform-dependent manner in murine models [166]. On the other hand, reduction in ApoE levels in mice models using anti-Apoe ε4 monoclonal antibody seemed to prevent cognitive impairment and brain hyperphosphorylation [167]. Recently, it was found that an anti-human ApoE antibody specifically recognizes human ApoE ε4 and ApoE ε3 and preferentially binds nonlipidated, aggregated ApoE in mouse models expressing human ApoE and human Aβ [168]. Other therapeutic approaches look at modifying ApoE properties through structural modification, an increase in ApoE lipidation, or blocking its interaction. CRISPR/Cas9 has been used to transform ApoE ε4 into ApoE ε3 in mouse astrocytes [169,170]. Recently, Wang et al. (2018) was successful at converting ApoE ε4 to ApoE ε3 in iPSC-derived neurons and proved that the introduction of ApoE ε4 recapitulated the pathogenic effects [53].

Given the importance of the immune response in the pathology of AD, therapies targeting this process, mostly through *CD33* and *TREM2*, are moving into the clinical trial phase, as announced at the 14th International Conference on Alzheimer's and Parkinson's Diseases, held 27–31 March in Lisbon, Portugal. In particular, two groups, the biotech Alector, Inc. and the German Center for Neurodegenerative Diseases in Munich, have developed antibodies that activate TREM2. These antibodies will trigger signaling through its co-receptor DAP12 resulting in phosphorylation of Syk and the downstream activation of microglia to remove amyloid. The Alector antibody (AL002) has moved into phase I clinical study. Similarly, Alector has started its clinical trial of the anti-CD33 antibody (AL003). Taking into account the time-specific protective vs. harmful effect that microglia have in AD, if these antibodies work as expected, they would need to be administered at very specific time points.

Finally, other features of age-related diseases are BBB integrity and the accumulation of senescent cells. BBB integrity is essential for the (i) Aβ-clearance and (ii) lipid transport. Docosahexaenoid acid (DHA) is a blood-based essential fatty acid for cognition, and current clinical trials are looking at the cognitive benefits of taking DHA diet supplements. Pan et al. (2016) showed reduced DHA levels and cognitive response in fatty acid-binding protein 5 (FABP5) knockout mice, suggesting that FABP5 upregulation could be an alternative approach to improve DHA uptake and rescue cognitive function [171]. Zang et al. (2019) studied the brains of patients with AD and the transgenic APP/PS1 mouse model of AD. They observed that oligodendrocyte progenitor cells (OPC—brain cells mobilized in response to neuronal injury and demyelination) accumulate around Aβ plaques and acquire a senescent phenotype characterized by the upregulation of p21/CDKN1A and p16/INK4/CDKN2A proteins and β-galactosidase activity. They observed that senolytic treatment (dasatinib plus quercetin) improved APP/PS1 AD mouse model condition by removing p16-expressing OPCs from Aβ plaques (after 9 days of treatment), reducing Aβ-plaque-associated proinflammatory cytokines and microglial activation, and reducing levels of inflammation and Aβ plaque size (after 11 weeks of treatment). Altogether, senolytic treatment improved the hippocampus-dependent learning and memory capabilities of APP/PS1 AD mice [172]. Quercetin is a flavonoid with antioxidant and anti-inflammatory effects found in many plants and foods such as berries, green tea, and Ginko biloba, among others; natural products have the benefit of being readily available, as such some of them are being tested in animal models, for their neuroprotective, anti-inflammatory, antioxidant, anti-amyloidogenic, anticholinesterase properties, as potential therapeutics for AD [173,174].

5. Conclusions and Future Directions

The recent studies of the molecular mechanisms of AD have shown us that amyloid accumulation does not only trigger tau hyperphosphorylation and immune response, but it starts other series of events that contribute to increased stress in the brain—e.g., reduction in brain blood flow or increment of neuronal hyperactivity. In addition, Aβ activates the inflammasome and p38MAPK pathway, which stimulates the production of cytokines that promote tau hyperphosphorylation. Once the pathological environment is started, *APOE* can exacerbate the situation, both through the Aβ and tau pathways in an isoform-dependent manner, but *APOE*, in turn, seems to be regulated by *TREM2*. All of the different molecular mechanisms are highly interconnected and participate in AD pathogenesis at different time points. Therefore, future research should focus on identifying the potential triggers for non-ADAD etiologies, whether it is searching for additional rare coding variants in the many loci associated with the disease or exploring the non-coding regions of the genome for downstream effects modifying gene expression. Additionally, improvements are needed in the in vitro systems used to study the disease since we have seen that the response in mouse models differs from that of humans. By better understanding the chain of pathologic events associated with different genetic risk factors, we can potentially identify AD subtypes related to specific genetic architectures allowing for personalized diagnosis and treatments. So far, our capacity to predict AD is quite limited, with PRS having a prediction accuracy between 65% and 75%. Our tools for early detection are limited as well since we cannot currently detect individuals at risk until their Aβ and tau load has already built up. Ultimately, this is a detriment to developing and testing novel therapies in the right groups of participants.

The progress we have made in recent years in the understanding of AD has been monumental. Yet, there is still substantial work to do before we fully understand and control this disease. The generation of larger genetic studies and incorporation of rare variants in prediction models will facilitate the development of improved PRS for the prediction of the baseline risk of developing AD and will also allow for the identification of potential AD subtypes. In addition, the discovery of dynamic biomarkers will enable the prediction of age at onset and the rate of progression of the disease. Omic approaches can facilitate progress in this area by exploring changes in the proteomic and metabolomic profiles of individuals at different time points. Finally, to improve and reach a personalized medicine for AD, future studies need to incorporate ethnic diversity in the recruitment process as modeling of this disease has, so far, been almost exclusively done with European and American populations of Caucasian background.

Author Contributions: M.V.F. and C.C. designed the structure of the manuscript. M.V.F. was a major contributor in writing the manuscript, L.I. contributed to the biomarkers and early detection and revised the manuscript, and C.C. revised, read, and approved the manuscript. All authors have read and agreed to the published version of the manuscript.

Funding: This work was supported by grants from the NIH to C.C. (R01AG044546, P01AG003991, RF1AG053303, RF1AG058501, U01AG058922), M.V.F. (K99AG061281), and L.I. (K99AG062723). C.C. was also supported by the Alzheimer's Association (NIRG-11-200110, BAND-14-338165, AARG-16-441560, and BFG-15-362540). L.I. was also supported by the Alzheimer Drug Discovery Foundation (GDAPB2018072015632) and the U.S. Department of Defense (W81XWH2010849).

Acknowledgments: We thank all participants and their families for their commitment and dedication to helping advance research into the early detection and causation of AD. Figures for this paper were created with BioRender.com.

Conflicts of Interest: C.C. receives research support from: Biogen, EISAI, Alector, and Parabon. The funders of the study had no role in the collection, analysis, or interpretation of data; in the writing of the report; or in the decision to submit the paper for publication. C.C. is a member of the advisory board of Vivid Genetics, Halia Therapeutics, and ADx Healthcare.

Abbreviations

5xFAD	Transgenic mouse model carrying APP KM670/671NL (Swedish), APP I716V (Florida), APP V717I (London), PSEN1 M146L (A>C), PSEN1 L286V mutations
AAO	Age at onset
ABCA7	ATP-binding cassette subfamily A member 7
ACE	Angiotensin I-converting enzyme
AD	Alzheimer's disease
ADAD	Autosomal dominant Alzheimer's disease
ADAM10	ADAM metallopeptidase domain 10
ALS	Amyotrohpic lateral sclerosis
ANK1	Ankyrin 1
APOE	Apolipoprotein E
APP	Amyloid precursor protein
APP^{NL-G-F}	Knock-in mice model carrying the APP KM670/671NL (Swedish), APP I716F (Iberian), APP E693G (Arctic) mutations
APPswe/PS1dE9	Transgenic mouse model carrying APP KM670/671NL (Swedish), PSEN1: deltaE9 mutations

APPPS1-21	Transgenic mouse model carrying APP KM670/671NL (Swedish), PSEN1 L166P
ASO	Antisense oligonucleotides
Aβ	Amyloid β
BACE	β secretase, also known as β-site Amyloid precursor protein Ceaving Enzyme
BBB	Blood-brain barrier
BV2	Immortalized murine microglial cell line
CD2AP	CD2-associated protein
CD33	Myeloid cell surface antigen CD33
CDR	Clinical dementia ratio
CE	Cholesterol ester
CERAD	Consortium to Establish a Registry for Alzheimer's Disease
CELF1	CUGBP Elav-like family member 1
CLU	Clusterin
CNS	Central nervous system
CNV	Copy number variant
CR1	Complement C3b/C4b receptor 1 (Knops blood group)
CSF	Cerebrospinal fluid
CT	Computerized tomography
DAM	Disease-associated microglia
DIAN	Dominantly inherited network
DLB	Dementia with Levy bodies
EMP	Erithro-myeloid progenitor
EAAT	Excitatory amino acid-mediated transporters
EOAD	Early onset Alzheimer's disease
EPHA1	EPH receptor A1
ERK	Extracellular signal-regulated kinase
ET_A	Endothelin A receptor
EWAs	Epigenome-wide association studies
fAD	Familial Alzheimer's disease
fLOAD	Familial late-onset Alzheimer's disease
FDR	False discovery rate
FTD	Frontotemporal dementia
GWA	Genome-wide association
H3K9ac	Histone 3 lysine 9 acetylation
H4K16ac	Histone 4 lysine 16 acetylation
HD	Huntington's disease
HSPG	Heparan sulfate proteoglycan
iPSC	Induced pluripotent stem cells
ITAM	Immunoreceptor tyrosine-based activation motif
LDLR	Low-density lipoprotein receptor
LOAD	Late-onset Alzheimer's disease
LRP1	LDL (low-density lipoprotein) receptor-related protein 1
MAP	Memory and aging project
MAPK	Mitogen-activated protein kinases (formerly known as ERK)
MAPT	Microtubule-associated protein tau
MCI	Mild cognitive impairment
MDMi	Monocyte-derived microglia-like cells
MKK7	MAPK kinase 7
MMSE	Mini-mental state examination
MOca	Montreal Cognitive Assessment
MRI	Magnetic resonance imaging
MS4A	Membrane spanning 4
MS4A4A	Membrane spanning 4-domains A4A
mTOR	Mechanistic target of rapamycin
NfL	Neurofilament Ligtht
NFTs	Neuro fibrillary tangles
NGRN	Neurogranin
NLRP3	Nucleotide-binding domain (NOD)-like receptor protein 3
NOX4	NADPH oxidase 4

NT	Neuropil threads
p38MAPK	p38 mitogen-activated protein kinase
PD	Parkinson's disease
PET	Positron emission tomography
PGRN	Progranulin
PI3K	Phosphatidylinositol-3-kinase
PIB	Pittsburgh compound B
PICALM	Phosphatidylinositol-binding clathrin assembly protein
PPAR	Peroxisome proliferator-activated receptor
PRS	Polygenic Risk Score
PS19	Transgenic mouse model carrying MAPT P301S mutation
PSEN1	Presenilin 1
PSEN2	Presenilin 2
ROS	Religious Order Study
sAD	Sporadic Alzheimer's disease
sLOAD	Sporadic late-onset Alzheimer's disease
SNPs	Single nucleotide polymorphism
sPDGFRβ	Soluble platelet-derived growth factor receptor-β
SPI1	Spi-1 proto-oncogene
TLRs	Toll-like receptors
TMEM106B	Transmembrane protein 106B
TREM2	Triggering receptor expressed on myeloid cells 2
TWAs	Transcriptome-wide association studies
TYROBP	Transmembrane immune signaling adaptor TYROBP
VD	Vascular dementia
VILIP1	Visinin-like 1

References

1. Livingston, G.; Sommerlad, A.; Orgeta, V.; Costafreda, S.G.; Huntley, J.; Ames, D.; Ballard, C.; Banerjee, S.; Burns, A.; Cohen-Mansfield, J.; et al. Dementia prevention, intervention, and care. *Lancet* **2017**, *390*, 2673–2734. [CrossRef]
2. Alzheimer's Association. 2019 Alzheimer's Disease Facts and Figures. In *Alzheimer's & Dementia*; Alzheimer's Association: Chicago, IL, USA, 2019; p. 87.
3. Hardy, J.A.; Higgins, G.A. Alzheimer's disease: The amyloid cascade hypothesis. *Science* **1992**, *256*, 184–185. [CrossRef]
4. Lambert, J.C.; Ibrahim-Verbaas, C.A.; Harold, D.; Naj, A.C.; Sims, R.; Bellenguez, C.; DeStafano, A.L.; Bis, J.C.; Beecham, G.W.; Grenier-Boley, B.; et al. Meta-analysis of 74,046 individuals identifies 11 new susceptibility loci for Alzheimer's disease. *Nat. Genet.* **2013**, *45*, 1452–1458. [CrossRef]
5. Sims, R.; Ibrahim-Verbaas, C.A.; Harold, D.; Naj, A.C.; Sims, R.; Bellenguez, C.; DeStafano, A.L.; Bis, J.C.; Beecham, G.W.; Grenier-Boley, B.; et al. Rare coding variants in PLCG2, ABI3, and TREM2 implicate microglial-mediated innate immunity in Alzheimer's disease. *Nat. Genet.* **2017**, *49*, 1373–1384. [CrossRef]
6. Jansen, I.E.; Savage, J.E.; Watanabe, K.; Bryois, J.; Williams, D.M.; Steinberg, S.; Sealock, J.; Karlsson, I.K.; Hagg, S.; Athanasiu, L.; et al. Genome-wide meta-analysis identifies new loci and functional pathways influencing Alzheimer's disease risk. *Nat. Genet.* **2019**, *51*, 404–413. [CrossRef]
7. Kunkle, B.W.; Grenier-Boley, B.; Sims, R.; Bis, J.C.; Damotte, V.; Naj, A.C.; Boland, A.; Vronskaya, M.; van der Lee, S.J.; Amlie-Wolf, A.; et al. Genetic meta-analysis of diagnosed Alzheimer's disease identifies new risk loci and implicates Abeta, tau, immunity and lipid processing. *Nat. Genet.* **2019**, *51*, 414–430. [CrossRef] [PubMed]
8. Ulrich, J.D.; Ulland, T.K.; Mahan, T.E.; Nystrom, S.; Nilsson, K.P.; Song, W.M.; Zhou, Y.; Reinartz, M.; Choi, S.; Jiang, H.; et al. ApoE facilitates the microglial response to amyloid plaque pathology. *J. Exp. Med.* **2018**, *215*, 1047–1058. [CrossRef]
9. Novikova, G.; Kapoor, M.; Tcw, J.; Abud, E.M.; Efthymiou, A.G.; Chen, S.X.; Cheng, H.; Fullard, J.F.; Bendl, J.; Liu, Y.; et al. Integration of Alzheimer's disease genetics and myeloid genomics identifies disease risk regulatory elements and genes. *Nat. Commun.* **2021**, *12*, 1610. [CrossRef] [PubMed]
10. Hernandez, I.; Mauleon, A.; Rosense-Roca, M.; Alegret, M.; Vinyes, G.; Espinosa, A.; Sotolongo-Grau, O.; Becker, J.T.; Valero, S.; Tarraga, L.; et al. Identification of misdiagnosed fronto-temporal dementia using APOE genotype and phenotype-genotype correlation analyses. *Curr. Alzheimer Res.* **2014**, *11*, 182–191. [CrossRef]
11. Landau, S.M.; Horng, A.; Fero, A.; Jagust, W.J. Amyloid negativity in patients with clinically diagnosed Alzheimer disease and MCI. *Neurology* **2016**, *86*, 1377–1385. [CrossRef]
12. van Dyck, C.H. Anti-Amyloid-β Monoclonal Antibodies for Alzheimer's Disease: Pitfalls and Promise. *Biol. Psychiatry* **2018**, *83*, 311–319. [CrossRef]
13. Cummings, J.; Ritter, A.; Zhong, K. Clinical Trials for Disease-Modifying Therapies in Alzheimer's Disease: A Primer, Lessons Learned, and a Blueprint for the Future. *J. Alzheimers Dis.* **2018**, *64* (Suppl. S1), S3–S22. [CrossRef]

14. Holtzman, D.M.; Morris, J.C.; Goate, A.M. Alzheimer's disease: The challenge of the second century. *Sci. Transl. Med.* **2011**, *3*, 77sr1. [CrossRef] [PubMed]
15. Hardy, J.; Selkoe, D.J. The amyloid hypothesis of Alzheimer's disease: Progress and problems on the road to therapeutics. *Science* **2002**, *297*, 353–356. [CrossRef]
16. Gomez-Isla, T.; Hollister, R.; West, H.; Mui, S.; Growdon, J.H.; Petersen, R.C.; Parisi, J.E.; Hyman, B.T. Neuronal loss correlates with but exceeds neurofibrillary tangles in Alzheimer's disease. *Ann. Neurol.* **1997**, *41*, 17–24. [CrossRef] [PubMed]
17. Gomez-Isla, T.; Hollister, R.; West, H.; Mui, S.; Growdon, J.H.; Petersen, R.C.; Parisi, J.E.; Hyman, B.T. Neurofibrillary tangles mediate the association of amyloid load with clinical Alzheimer disease and level of cognitive function. *Arch. Neurol.* **2004**, *61*, 378–384.
18. Bejanin, A.; Schonhaut, D.R.; La Joie, R.; Kramer, J.H.; Baker, S.L.; Sosa, N.; Ayakta, N.; Cantwell, A.; Janabi, M.; Lauriola, M.; et al. Tau pathology and neurodegeneration contribute to cognitive impairment in Alzheimer's disease. *Brain* **2017**, *140*, 3286–3300. [CrossRef]
19. Nortley, R.; Korte, N.; Izquierdo, P.; Hirunpattarasilp, C.; Mishra, A.; Jaunmuktane, Z.; Kyrargyri, V.; Pfeiffer, T.; Khennouf, L.; Madry, C.; et al. Amyloid β oligomers constrict human capillaries in Alzheimer's disease via signaling to pericytes. *Science* **2019**, *365*, eaav9518. [CrossRef]
20. Zott, B.; Simon, M.M.; Hong, W.; Unger, F.; Chen-Engerer, H.J.; Frosch, M.P.; Sakmann, B.; Walsh, D.M.; Konnerth, A. A vicious cycle of β amyloid-dependent neuronal hyperactivation. *Science* **2019**, *365*, 559–565. [CrossRef] [PubMed]
21. Venegas, C.; Kumar, S.; Franklin, B.S.; Dierkes, T.; Brinkschulte, R.; Tejera, D.; Vieira-Saecker, A.; Schwartz, S.; Santarelli, F.; Kummer, M.P.; et al. Microglia-derived ASC specks cross-seed amyloid-β in Alzheimer's disease. *Nature* **2017**, *552*, 355–361. [CrossRef] [PubMed]
22. Ising, C.; Venegas, C.; Zhang, S.; Scheiblich, H.; Schmidt, S.V.; Vieira-Saecker, A.; Schwartz, S.; Albasset, S.; McManus, R.M.; Tejera, D.; et al. NLRP3 inflammasome activation drives tau pathology. *Nature* **2019**, *575*, 669–673. [CrossRef]
23. He, Z.; Guo, J.L.; McBride, J.D.; Narasimhan, S.; Kim, H.; Changolkar, L.; Zhang, B.; Gathagan, R.J.; Yue, C.; Dengler, C.; et al. Amyloid-β plaques enhance Alzheimer's brain tau-seeded pathologies by facilitating neuritic plaque tau aggregation. *Nat. Med.* **2018**, *24*, 29–38. [CrossRef]
24. Maphis, N.; Jiang, S.; Xu, G.; Kokiko-Cochran, O.N.; Roy, S.M.; Van Eldik, L.J.; Watterson, D.M.; Lamb, B.T.; Bhaskar, K. Selective suppression of the α isoform of p38 MAPK rescues late-stage tau pathology. *Alzheimers Res. Ther.* **2016**, *8*, 54. [CrossRef]
25. Ittner, A.; Chua, S.W.; Bertz, J.; Volkerling, A.; van der Hoven, J.; Gladbach, A.; Przybyla, M.; Bi, M.; van Hummel, A.; Stevens, C.H.; et al. Site-specific phosphorylation of tau inhibits amyloid-β toxicity in Alzheimer's mice. *Science* **2016**, *354*, 904–908. [CrossRef] [PubMed]
26. Mass, E.; Jacome-Galarza, C.E.; Blank, T.; Lazarov, T.; Durham, B.H.; Ozkaya, N.; Pastore, A.; Schwabenland, M.; Chung, Y.R.; Rosenblum, M.K.; et al. A somatic mutation in erythro-myeloid progenitors causes neurodegenerative disease. *Nature* **2017**, *549*, 389–393. [CrossRef]
27. Klein, H.U.; McCabe, C.; Gjoneska, E.; Sullivan, S.E.; Kaskow, B.J.; Tang, A.; Smith, R.V.; Xu, J.; Pfenning, A.R.; Bernstein, B.E.; et al. Epigenome-wide study uncovers large-scale changes in histone acetylation driven by tau pathology in aging and Alzheimer's human brains. *Nat. Neurosci.* **2019**, *22*, 37–46. [CrossRef]
28. Getz, G.S.; Reardon, C.A. Apoprotein E as a lipid transport and signaling protein in the blood, liver, and artery wall. *J. Lipid Res.* **2009**, *50*, S156–S161. [CrossRef]
29. Farrer, L.A.; Cupples, L.A.; Haines, J.L.; Hyman, B.; Kukull, W.A.; Mayeux, R.; Myers, R.H.; Pericak-Vance, M.A.; Risch, N.; van Duijn, C.M. Effects of age, sex, and ethnicity on the association between apolipoprotein E genotype and Alzheimer disease. A meta-analysis. APOE and Alzheimer Disease Meta Analysis Consortium. *JAMA* **1997**, *278*, 1349–1356. [CrossRef] [PubMed]
30. Frieden, C.; Garai, K. Structural differences between apoE3 and apoE4 may be useful in developing therapeutic agents for Alzheimer's disease. *Proc. Natl. Acad. Sci. USA* **2012**, *109*, 8913–8918. [CrossRef] [PubMed]
31. Zhong, N.; Weisgraber, K.H. Understanding the association of apolipoprotein E4 with Alzheimer disease: Clues from its structure. *J. Biol. Chem.* **2009**, *284*, 6027–6031. [CrossRef]
32. Holtzman, D.M.; Herz, J.; Bu, G. Apolipoprotein E and apolipoprotein E receptors: Normal biology and roles in Alzheimer disease. *Cold Spring Harb. Perspect. Med.* **2012**, *2*, a006312. [CrossRef]
33. Mahley, R.W.; Rall, S.C., Jr. Apolipoprotein E: Far more than a lipid transport protein. *Annu. Rev. Genom. Hum. Genet.* **2000**, *1*, 507–537. [CrossRef]
34. Mahley, R.W.; Rall, S.C., Jr. Apolipoprotein E; Apolipoprotein E genotype and cardiovascular disease in the Framingham Heart Study. *Atherosclerosis* **2001**, *154*, 529–537.
35. Slooter, A.J.; Cruts, M.; Kalmijn, S.; Hofman, A.; Breteler, M.M.; Van Broeckhoven, C.; van Duijn, C.M. Risk estimates of dementia by apolipoprotein E genotypes from a population-based incidence study: The Rotterdam Study. *Arch. Neurol.* **1998**, *55*, 964–968. [CrossRef] [PubMed]
36. Maestre, G.; Ottman, R.; Stern, Y.; Gurland, B.; Chun, M.; Tang, M.X.; Shelanski, M.; Tycko, B.; Mayeux, R. Apolipoprotein E and Alzheimer's disease: Ethnic variation in genotypic risks. *Ann. Neurol.* **1995**, *37*, 254–259. [CrossRef] [PubMed]
37. van der Lee, S.J.; Wolters, F.J.; Ikram, M.K.; Hofman, A.; Ikram, M.A.; Amin, N.; van Duijn, C.M. The effect of APOE and other common genetic variants on the onset of Alzheimer's disease and dementia: A community-based cohort study. *Lancet Neurol.* **2018**, *17*, 434–444. [CrossRef]

38. Pastor, P.; Roe, C.M.; Villegas, A.; Bedoya, G.; Chakraverty, S.; Garcia, G.; Tirado, V.; Norton, J.; Rios, S.; Martinez, M.; et al. Apolipoprotein Eepsilon4 modifies Alzheimer's disease onset in an E280A PS1 kindred. *Ann. Neurol.* **2003**, *54*, 163–169. [CrossRef]
39. Corder, E.H.; Saunders, A.M.; Strittmatter, W.J.; Schmechel, D.E.; Gaskell, P.C.; Small, G.W.; Roses, A.D.; Haines, J.L.; Pericak-Vance, M.A. Gene dose of apolipoprotein E type 4 allele and the risk of Alzheimer's disease in late onset families. *Science* **1993**, *261*, 921–923. [CrossRef] [PubMed]
40. Corder, E.H.; Saunders, A.M.; Risch, N.J.; Strittmatter, W.J.; Schmechel, D.E.; Gaskell, P.C., Jr.; Rimmler, J.B.; Locke, P.A.; Conneally, P.M.; Schmader, K.E.; et al. Protective effect of apolipoprotein E type 2 allele for late onset Alzheimer disease. *Nat. Genet.* **1994**, *7*, 180–184. [CrossRef]
41. Arboleda-Velasquez, J.F.; Lopera, F.; O'Hare, M.; Delgado-Tirado, S.; Marino, C.; Chmielewska, N.; Saez-Torres, K.L.; Amarnani, D.; Schultz, A.P.; Sperling, R.A.; et al. Resistance to autosomal dominant Alzheimer's disease in an APOE3 Christchurch homozygote: A case report. *Nat. Med.* **2019**, *25*, 1680–1683. [CrossRef]
42. Hashimoto, T.; Serrano-Pozo, A.; Hori, Y.; Adams, K.W.; Takeda, S.; Banerji, A.O.; Mitani, A.; Joyner, D.; Thyssen, D.H.; Bacskai, B.J.; et al. Apolipoprotein E, especially apolipoprotein E4, increases the oligomerization of amyloid β peptide. *J. Neurosci.* **2012**, *32*, 15181–15192. [CrossRef] [PubMed]
43. Koffie, R.M.; Hashimoto, T.; Tai, H.C.; Kay, K.R.; Serrano-Pozo, A.; Joyner, D.; Hou, S.; Kopeikina, K.J.; Frosch, M.P.; Lee, V.M.; et al. Apolipoprotein E4 effects in Alzheimer's disease are mediated by synaptotoxic oligomeric amyloid-β. *Brain* **2012**, *135*, 2155–2168. [CrossRef]
44. Christensen, D.Z.; Schneider-Axmann, T.; Lucassen, P.J.; Bayer, T.A.; Wirths, O. Accumulation of intraneuronal Abeta correlates with ApoE4 genotype. *Acta Neuropathol.* **2010**, *119*, 555–566. [CrossRef] [PubMed]
45. Kanekiyo, T.; Zhang, J.; Liu, Q.; Liu, C.C.; Zhang, L.; Bu, G. Heparan sulphate proteoglycan and the low-density lipoprotein receptor-related protein 1 constitute major pathways for neuronal amyloid-β uptake. *J. Neurosci.* **2011**, *31*, 1644–1651. [CrossRef] [PubMed]
46. Kim, J.; Castellano, J.M.; Jiang, H.; Basak, J.M.; Parsadanian, M.; Pham, V.; Mason, S.M.; Paul, S.M.; Holtzman, D.M. Overexpression of low-density lipoprotein receptor in the brain markedly inhibits amyloid deposition and increases extracellular Aβ clearance. *Neuron* **2009**, *64*, 632–644. [CrossRef]
47. Bu, G. Apolipoprotein E and its receptors in Alzheimer's disease: Pathways, pathogenesis and therapy. *Nat. Rev. Neurosci.* **2009**, *10*, 333–344. [CrossRef]
48. Huang, Y.A.; Zhou, B.; Wernig, M.; Sudhof, T.C. ApoE2, ApoE3, and ApoE4 Differentially Stimulate APP Transcription and Abeta Secretion. *Cell* **2017**, *168*, 427–441.e21. [CrossRef]
49. Cruchaga, C.; Kauwe, J.S.; Harari, O.; Jin, S.C.; Cai, Y.; Karch, C.M.; Benitez, B.A.; Jeng, A.T.; Skorupa, T.; Carrell, D.; et al. GWAS of cerebrospinal fluid tau levels identifies risk variants for Alzheimer's disease. *Neuron* **2013**, *78*, 256–268. [CrossRef]
50. Kauwe, J.S.; Wang, J.; Mayo, K.; Morris, J.C.; Fagan, A.M.; Holtzman, D.M.; Goate, A.M. Alzheimer's disease risk variants show association with cerebrospinal fluid amyloid β. *Neurogenetics* **2009**, *10*, 13–17. [CrossRef]
51. Deming, Y.; Li, Z.; Kapoor, M.; Harari, O.; Del-Aguila, J.L.; Black, K.; Carrell, D.; Cai, Y.; Fernandez, M.V.; Budde, J.; et al. Genome-wide association study identifies four novel loci associated with Alzheimer's endophenotypes and disease modifiers. *Acta Neuropathol.* **2017**, *133*, 839–856. [CrossRef]
52. Shi, Y.; Yamada, K.; Liddelow, S.A.; Smith, S.T.; Zhao, L.; Luo, W.; Tsai, R.M.; Spina, S.; Grinberg, L.T.; Rojas, J.C.; et al. ApoE4 markedly exacerbates tau-mediated neurodegeneration in a mouse model of tauopathy. *Nature* **2017**, *549*, 523–527. [CrossRef] [PubMed]
53. Wang, C.; Najm, R.; Xu, Q.; Jeong, D.E.; Walker, D.; Balestra, M.E.; Yoon, S.Y.; Yuan, H.; Li, G.; Miller, Z.A.; et al. Gain of toxic apolipoprotein E4 effects in human iPSC-derived neurons is ameliorated by a small-molecule structure corrector. *Nat. Med.* **2018**, *24*, 647–657. [CrossRef]
54. Pimenova, A.A.; Marcora, E.; Goate, A.M. A Tale of Two Genes: Microglial Apoe and Trem2. *Immunity* **2017**, *47*, 398–400. [CrossRef]
55. Krasemann, S.; Madore, C.; Cialic, R.; Baufeld, C.; Calcagno, N.; El Fatimy, R.; Beckers, L.; O'Loughlin, E.; Xu, Y.; Fanek, Z.; et al. The TREM2-APOE Pathway Drives the Transcriptional Phenotype of Dysfunctional Microglia in Neurodegenerative Diseases. *Immunity* **2017**, *47*, 566–581.e9. [CrossRef] [PubMed]
56. Parhizkar, S.; Arzberger, T.; Brendel, M.; Kleinberger, G.; Deussing, M.; Focke, C.; Nuscher, B.; Xiong, M.; Ghasemigharagoz, A.; Katzmarski, N.; et al. Loss of TREM2 function increases amyloid seeding but reduces plaque-associated ApoE. *Nat. Neurosci.* **2019**, *22*, 191–204. [CrossRef]
57. Nugent, A.A.; Lin, K.; van Lengerich, B.; Lianoglou, S.; Przybyla, L.; Davis, S.S.; Llapashtica, C.; Wang, J.; Kim, D.J.; Xia, D.; et al. TREM2 Regulates Microglial Cholesterol Metabolism upon Chronic Phagocytic Challenge. *Neuron* **2020**, *105*, 837–854.e9. [CrossRef]
58. Bertram, L.; Lange, C.; Mullin, K.; Parkinson, M.; Hsiao, M.; Hogan, M.F.; Schjeide, B.M.; Hooli, B.; Divito, J.; Ionita, I.; et al. Genome-wide association analysis reveals putative Alzheimer's disease susceptibility loci in addition to APOE. *Am. J. Hum. Genet.* **2008**, *83*, 623–632. [CrossRef] [PubMed]
59. Harold, D.; Abraham, R.; Hollingworth, P.; Sims, R.; Gerrish, A.; Hamshere, M.L.; Pahwa, J.S.; Moskvina, V.; Dowzell, K.; Williams, A.; et al. Genome-wide association study identifies variants at CLU and PICALM associated with Alzheimer's disease. *Nat. Genet.* **2009**, *41*, 1088–1093. [CrossRef] [PubMed]

60. Seshadri, S.; Fitzpatrick, A.L.; Ikram, M.A.; DeStefano, A.L.; Gudnason, V.; Boada, M.; Bis, J.C.; Smith, A.V.; Carassquillo, M.M.; Lambert, J.C.; et al. Genome-wide analysis of genetic loci associated with Alzheimer disease. *JAMA* **2010**, *303*, 1832–1840. [CrossRef]
61. Hollingworth, P.; Harold, D.; Sims, R.; Gerrish, A.; Lambert, J.C.; Carrasquillo, M.M.; Abraham, R.; Hamshere, M.L.; Pahwa, J.S.; Moskvina, V.; et al. Common variants at ABCA7, MS4A6A/MS4A4E, EPHA1, CD33 and CD2AP are associated with Alzheimer's disease. *Nat. Genet.* **2011**, *43*, 429–435. [CrossRef]
62. Naj, A.C.; Jun, G.; Beecham, G.W.; Wang, L.S.; Vardarajan, B.N.; Buros, J.; Gallins, P.J.; Buxbaum, J.D.; Jarvik, G.P.; Crane, P.K.; et al. Common variants at MS4A4/MS4A6E, CD2AP, CD33 and EPHA1 are associated with late-onset Alzheimer's disease. *Nat. Genet.* **2011**, *43*, 436–441. [CrossRef]
63. Benitez, B.A.; Jin, S.C.; Guerreiro, R.; Graham, R.; Lord, J.; Harold, D.; Sims, R.; Lambert, J.C.; Gibbs, J.R.; Bras, J.; et al. Missense variant in TREML2 protects against Alzheimer's disease. *Neurobiol. Aging* **2014**, *35*, 1510.e19. [CrossRef] [PubMed]
64. Guerreiro, R.J.; Lohmann, E.; Bras, J.M.; Gibbs, J.R.; Rohrer, J.D.; Gurunlian, N.; Dursun, B.; Bilgic, B.; Hanagasi, H.; Gurvit, H.; et al. Using exome sequencing to reveal mutations in TREM2 presenting as a frontotemporal dementia-like syndrome without bone involvement. *JAMA Neurol.* **2013**, *70*, 78–84. [CrossRef] [PubMed]
65. Huang, K.L.; Marcora, E.; Pimenova, A.A.; Di Narzo, A.F.; Kapoor, M.; Jin, S.C.; Harari, O.; Bertelsen, S.; Fairfax, B.P.; Czajkowski, J.; et al. A common haplotype lowers PU.1 expression in myeloid cells and delays onset of Alzheimer's disease. *Nat. Neurosci.* **2017**, *20*, 1052–1061. [CrossRef] [PubMed]
66. Manzine, P.R.; Ettcheto, M.; Cano, A.; Busquets, O.; Marcello, E.; Pelucchi, S.; Di Luca, M.; Endres, K.; Olloquequi, J.; Camins, A.; et al. ADAM10 in Alzheimer's disease: Pharmacological modulation by natural compounds and its role as a peripheral marker. *Biomed. Pharmacother.* **2019**, *113*, 108661. [CrossRef]
67. Jochemsen, H.M.; Teunissen, C.E.; Ashby, E.L.; van der Flier, W.M.; Jones, R.E.; Geerlings, M.I.; Scheltens, P.; Kehoe, P.G.; Muller, M. The association of angiotensin-converting enzyme with biomarkers for Alzheimer's disease. *Alzheimers Res. Ther.* **2014**, *6*, 27. [CrossRef]
68. Koronyo-Hamaoui, M.; Sheyn, J.; Hayden, E.Y.; Li, S.; Fuchs, D.T.; Regis, G.C.; Lopes, D.H.J.; Black, K.L.; Bernstein, K.E.; Teplow, D.B.; et al. Peripherally derived angiotensin converting enzyme-enhanced macrophages alleviate Alzheimer-related disease. *Brain* **2020**, *143*, 336–358. [CrossRef]
69. Wang, W.Y.; Tan, M.S.; Yu, J.T.; Tan, L. Role of pro-inflammatory cytokines released from microglia in Alzheimer's disease. *Ann. Transl. Med.* **2015**, *3*, 136.
70. Baik, S.H.; Kang, S.; Lee, W.; Choi, H.; Chung, S.; Kim, J.I.; Mook-Jung, I.A. A Breakdown in Metabolic Reprogramming Causes Microglia Dysfunction in Alzheimer's Disease. *Cell Metab.* **2019**, *30*, 493–507.e6. [CrossRef]
71. Piers, T.M.; Cosker, K.; Mallach, A.; Johnson, G.T.; Guerreiro, R.; Hardy, J.; Pocock, J.M. A locked immunometabolic switch underlies TREM2 R47H loss of function in human iPSC-derived microglia. *FASEB J.* **2020**, *34*, 2436–2450. [CrossRef]
72. Ulland, T.K.; Colonna, M. TREM2—A key player in microglial biology and Alzheimer disease. *Nat. Rev. Neurol.* **2018**, *14*, 667–675. [CrossRef]
73. Griciuc, A.; Serrano-Pozo, A.; Parrado, A.R.; Lesinski, A.N.; Asselin, C.N.; Mullin, K.; Hooli, B.; Choi, S.H.; Hyman, B.T.; Tanzi, R.E. Alzheimer's disease risk gene CD33 inhibits microglial uptake of amyloid β. *Neuron* **2013**, *78*, 631–643. [CrossRef] [PubMed]
74. Shi, Q.; Chowdhury, S.; Ma, R.; Le, K.X.; Hong, S.; Caldarone, B.J.; Stevens, B.; Lemere, C.A. Complement C3 deficiency protects against neurodegeneration in aged plaque-rich APP/PS1 mice. *Sci. Transl. Med.* **2017**, *9*, eaaf6295. [CrossRef] [PubMed]
75. Ewers, M.; Franzmeier, N.; Suarez-Calvet, M.; Morenas-Rodriguez, E.; Caballero, M.A.A.; Kleinberger, G.; Piccio, L.; Cruchaga, C.; Deming, Y.; Dichgans, M.; et al. Increased soluble TREM2 in cerebrospinal fluid is associated with reduced cognitive and clinical decline in Alzheimer's disease. *Sci. Transl. Med.* **2019**, *11*, eaav6221. [CrossRef]
76. Leyns, C.E.G.; Ulrich, J.D.; Finn, M.B.; Stewart, F.R.; Koscal, L.J.; Remolina Serrano, J.; Robinson, G.O.; Anderson, E.; Colonna, M.; Holtzman, D.M. TREM2 deficiency attenuates neuroinflammation and protects against neurodegeneration in a mouse model of tauopathy. *Proc. Natl. Acad. Sci. USA* **2017**, *114*, 11524–11529. [CrossRef] [PubMed]
77. Leyns, C.E.G.; Gratuze, M.; Narasimhan, S.; Jain, N.; Koscal, L.J.; Jiang, H.; Manis, M.; Colonna, M.; Lee, V.M.Y.; Ulrich, J.D.; et al. TREM2 function impedes tau seeding in neuritic plaques. *Nat. Neurosci.* **2019**, *22*, 1217–1222. [CrossRef]
78. Long, H.; Zhong, G.; Wang, C.; Zhang, J.; Zhang, Y.; Luo, J.; Shi, S. TREM2 Attenuates Abeta1-42-Mediated Neuroinflammation in BV-2 Cells by Downregulating TLR Signaling. *Neurochem. Res.* **2019**, *44*, 1830–1839. [CrossRef]
79. Deming, Y.; Li, Z.; Benitez, B.A.; Cruchaga, C. Triggering receptor expressed on myeloid cells 2 (TREM2): A potential therapeutic target for Alzheimer disease? *Expert Opin. Ther. Targets* **2018**, *22*, 587–598. [CrossRef]
80. Zhou, Y.; Song, W.M.; Andhey, P.S.; Swain, A.; Levy, T.; Miller, K.R.; Poliani, P.L.; Cominelli, M.; Grover, S.; Gilfillan, S.; et al. Human and mouse single-nucleus transcriptomics reveal TREM2-dependent and TREM2-independent cellular responses in Alzheimer's disease. *Nat. Med.* **2020**, *26*, 131–142. [CrossRef]
81. Deming, Y.; Filipello, F.; Cignarella, F.; Cantoni, C.; Hsu, S.; Mikesell, R.; Li, Z.; Del-Aguila, J.L.; Dube, U.; Farias, F.G.; et al. The MS4A gene cluster is a key modulator of soluble TREM2 and Alzheimer's disease risk. *Sci. Transl. Med.* **2019**, *11*, eaau2291. [CrossRef]
82. Kwart, D.; Gregg, A.; Scheckel, C.; Murphy, E.A.; Paquet, D.; Duffield, M.; Fak, J.; Olsen, O.; Darnell, R.B.; Tessier-Lavigne, M. A Large Panel of Isogenic APP and PSEN1 Mutant Human iPSC Neurons Reveals Shared Endosomal Abnormalities Mediated by APP β-CTFs, Not Aβ. *Neuron* **2019**, *104*, 256–270.e5. [CrossRef]
83. Van Acker, Z.P.; Bretou, M.; Annaert, W. Endo-lysosomal dysregulations and late-onset Alzheimer's disease: Impact of genetic risk factors. *Mol. Neurodegener.* **2019**, *14*, 20. [CrossRef]

84. Kwart, D.; Gregg, A.; Scheckel, C.; Murphy, E.A.; Paquet, D.; Duffield, M.; Fak, J.; Olsen, O.; Darnell, R.B.; Tessier-Lavigne, M.A. LC3-Associated Endocytosis Facilitates β-Amyloid Clearance and Mitigates Neurodegeneration in Murine Alzheimer's Disease. *Cell* 2019, *178*, 536–551.e14.
85. Jun, G.; Ibrahim-Verbaas, C.A.; Vronskaya, M.; Lambert, J.C.; Chung, J.; Naj, A.C.; Kunkle, B.W.; Wang, L.S.; Bis, J.C.; Bellenguez, C.; et al. A novel Alzheimer disease locus located near the gene encoding tau protein. *Mol. Psychiatry* 2016, *21*, 108–117. [CrossRef] [PubMed]
86. Li, Z.; Del-Aguila, J.L.; Dube, U.; Budde, J.; Martinez, R.; Black, K.; Xiao, Q.; Cairns, N.J.; Dominantly Inherited Alzheimer, N.; Dougherty, J.D.; et al. Genetic variants associated with Alzheimer's disease confer different cerebral cortex cell-type population structure. *Genome Med.* 2018, *10*, 43. [CrossRef]
87. Li, Z.; Farias, F.H.G.; Dube, U.; Del-Aguila, J.L.; Mihindukulasuriya, K.A.; Fernandez, M.V.; Ibanez, L.; Budde, J.P.; Wang, F.; Lake, A.M.; et al. The TMEM106B FTLD-protective variant, rs1990621, is also associated with increased neuronal proportion. *Acta Neuropathol.* 2020, *139*, 45–61. [CrossRef] [PubMed]
88. Lodato, M.A.; Rodin, R.E.; Bohrson, C.L.; Coulter, M.E.; Barton, A.R.; Kwon, M.; Sherman, M.A.; Vitzthum, C.M.; Luquette, L.J.; Yandava, C.N.; et al. Aging and neurodegeneration are associated with increased mutations in single human neurons. *Science* 2018, *359*, 555–559. [CrossRef] [PubMed]
89. Verheijen, B.M.; Vermulst, M.; van Leeuwen, F.W. Somatic mutations in neurons during aging and neurodegeneration. *Acta Neuropathol.* 2018, *135*, 811–826. [CrossRef] [PubMed]
90. Lee, J.H. Somatic mutations in disorders with disrupted brain connectivity. *Exp. Mol. Med.* 2016, *48*, e239. [CrossRef]
91. Bushman, D.M.; Kaeser, G.E.; Siddoway, B.; Westra, J.W.; Rivera, R.R.; Rehen, S.K.; Yung, Y.C.; Chun, J. Genomic mosaicism with increased amyloid precursor protein (APP) gene copy number in single neurons from sporadic Alzheimer's disease brains. *eLife* 2015, *4*, e05116. [CrossRef] [PubMed]
92. Nicolas, G.; Acuna-Hidalgo, R.; Keogh, M.J.; Quenez, O.; Steehouwer, M.; Lelieveld, S.; Rousseau, S.; Richard, A.C.; Oud, M.S.; Marguet, F.; et al. Somatic variants in autosomal dominant genes are a rare cause of sporadic Alzheimer's disease. *Alzheimers Dement.* 2018, *14*, 1632–1639. [CrossRef]
93. Annese, A.; Manzari, C.; Lionetti, C.; Picardi, E.; Horner, D.S.; Chiara, M.; Caratozzolo, M.F.; Tullo, A.; Fosso, B.; Pesole, G.; et al. Whole transcriptome profiling of Late-Onset Alzheimer's Disease patients provides insights into the molecular changes involved in the disease. *Sci. Rep.* 2018, *8*, 4282. [CrossRef]
94. Li, X.; Long, J.; He, T.; Belshaw, R.; Scott, J. Integrated genomic approaches identify major pathways and upstream regulators in late onset Alzheimer's disease. *Sci. Rep.* 2015, *5*, 12393. [CrossRef]
95. Wang, M.; Roussos, P.; McKenzie, A.; Zhou, X.; Kajiwara, Y.; Brennand, K.J.; De Luca, G.C.; Crary, J.F.; Casaccia, P.; Buxbaum, J.D.; et al. Integrative network analysis of nineteen brain regions identifies molecular signatures and networks underlying selective regional vulnerability to Alzheimer's disease. *Genome Med.* 2016, *8*, 104. [CrossRef]
96. Caberlotto, L.; Marchetti, L.; Lauria, M.; Scotti, M.; Parolo, S. Integration of transcriptomic and genomic data suggests candidate mechanisms for APOE4-mediated pathogenic action in Alzheimer's disease. *Sci. Rep.* 2016, *6*, 32583. [CrossRef]
97. Mostafavi, S.; Gaiteri, C.; Sullivan, S.E.; White, C.C.; Tasaki, S.; Xu, J.; Taga, M.; Klein, H.U.; Patrick, E.; Komashko, V.; et al. A molecular network of the aging human brain provides insights into the pathology and cognitive decline of Alzheimer's disease. *Nat. Neurosci.* 2018, *21*, 811–819. [CrossRef]
98. Sepulcre, J.; Grothe, M.J.; d'Oleire Uquillas, F.; Ortiz-Teran, L.; Diez, I.; Yang, H.S.; Jacobs, H.I.L.; Hanseeuw, B.J.; Li, Q.; El-Fakhri, G.; et al. Neurogenetic contributions to amyloid β and tau spreading in the human cortex. *Nat. Med.* 2018, *24*, 1910–1918. [CrossRef]
99. Del-Aguila, J.L.; Li, Z.; Dube, U.; Mihindukulasuriya, K.A.; Budde, J.P.; Fernandez, M.V.; Ibanez, L.; Bradley, J.; Wang, F.; Bergmann, K.; et al. A single-nuclei RNA sequencing study of Mendelian and sporadic AD in the human brain. *Alzheimers Res. Ther.* 2019, *11*, 71. [CrossRef] [PubMed]
100. Mathys, H.; Davila-Velderrain, J.; Peng, Z.; Gao, F.; Mohammadi, S.; Young, J.Z.; Menon, M.; He, L.; Abdurrob, F.; Jiang, X.; et al. Single-cell transcriptomic analysis of Alzheimer's disease. *Nature* 2019, *570*, 332–337. [CrossRef] [PubMed]
101. Srinivasan, K.; Friedman, B.A.; Etxeberria, A.; Huntley, M.A.; van der Brug, M.P.; Foreman, O.; Paw, J.S.; Modrusan, Z.; Beach, T.G.; Serrano, G.E.; et al. Alzheimer's Patient Microglia Exhibit Enhanced Aging and Unique Transcriptional Activation. *Cell Rep.* 2020, *31*, 107843. [CrossRef] [PubMed]
102. Dube, U.; Del-Aguila, J.L.; Li, Z.; Budde, J.P.; Jiang, S.; Hsu, S.; Ibanez, L.; Fernandez, M.V.; Farias, F.; Norton, J.; et al. An atlas of cortical circular RNA expression in Alzheimer disease brains demonstrates clinical and pathological associations. *Nat. Neurosci.* 2019, *22*, 1903–1912. [CrossRef]
103. De Jager, P.L.; Srivastava, G.; Lunnon, K.; Burgess, J.; Schalkwyk, L.C.; Yu, L.; Eaton, M.L.; Keenan, B.T.; Ernst, J.; McCabe, C.; et al. Alzheimer's disease: Early alterations in brain DNA methylation at ANK1, BIN1, RHBDF2 and other loci. *Nat. Neurosci.* 2014, *17*, 1156–1163. [CrossRef] [PubMed]
104. Lunnon, K.; Smith, R.; Hannon, E.; De Jager, P.L.; Srivastava, G.; Volta, M.; Troakes, C.; Al-Sarraj, S.; Burrage, J.; Macdonald, R.; et al. Methylomic profiling implicates cortical deregulation of ANK1 in Alzheimer's disease. *Nat. Neurosci.* 2014, *17*, 1164–1170. [CrossRef] [PubMed]
105. Gasparoni, G.; Bultmann, S.; Lutsik, P.; Kraus, T.F.J.; Sordon, S.; Vlcek, J.; Dietinger, V.; Steinmaurer, M.; Haider, M.; Mulholland, C.B.; et al. DNA methylation analysis on purified neurons and glia dissects age and Alzheimer's disease-specific changes in the human cortex. *Epigenet. Chromatin* 2018, *11*, 41. [CrossRef] [PubMed]

106. Smith, A.R.; Smith, R.G.; Burrage, J.; Troakes, C.; Al-Sarraj, S.; Kalaria, R.N.; Sloan, C.; Robinson, A.C.; Mill, J.; Lunnon, K. A cross-brain regions study of ANK1 DNA methylation in different neurodegenerative diseases. *Neurobiol. Aging* **2019**, *74*, 70–76. [CrossRef]
107. Nativio, R.; Donahue, G.; Berson, A.; Lan, Y.; Amlie-Wolf, A.; Tuzer, F.; Toledo, J.B.; Gosai, S.J.; Gregory, B.D.; Torres, C.; et al. Dysregulation of the epigenetic landscape of normal aging in Alzheimer's disease. *Nat. Neurosci.* **2018**, *21*, 497–505. [CrossRef]
108. Pagani, M.; Nobili, F.; Morbelli, S.; Arnaldi, D.; Giuliani, A.; Oberg, J.; Girtler, N.; Brugnolo, A.; Picco, A.; Bauckneht, M.; et al. Early identification of MCI converting to AD: A FDG PET study. *Eur. J. Nucl. Med. Mol. Imaging* **2017**, *44*, 2042–2052. [CrossRef]
109. Paglia, G.; Stocchero, M.; Cacciatore, S.; Lai, S.; Angel, P.; Alam, M.T.; Keller, M.; Ralser, M.; Astarita, G. Unbiased Metabolomic Investigation of Alzheimer's Disease Brain Points to Dysregulation of Mitochondrial Aspartate Metabolism. *J. Proteome Res.* **2016**, *15*, 608–618. [CrossRef]
110. Snowden, S.G.; Ebshiana, A.A.; Hye, A.; An, Y.; Pletnikova, O.; O'Brien, R.; Troncoso, J.; Legido-Quigley, C.; Thambisetty, M. Association between fatty acid metabolism in the brain and Alzheimer disease neuropathology and cognitive performance: A nontargeted metabolomic study. *PLoS Med.* **2017**, *14*, e1002266. [CrossRef]
111. Guiraud, S.P.; Montoliu, I.; Da Silva, L.; Dayon, L.; Galindo, A.N.; Corthesy, J.; Kussmann, M.; Martin, F.P. High-throughput and simultaneous quantitative analysis of homocysteine-methionine cycle metabolites and co-factors in blood plasma and cerebrospinal fluid by isotope dilution LC-MS/MS. *Anal. Bioanal. Chem.* **2017**, *409*, 295–305. [CrossRef]
112. Toledo, J.B.; Arnold, M.; Kastenmuller, G.; Chang, R.; Baillie, R.A.; Han, X.; Thambisetty, M.; Tenenbaum, J.D.; Suhre, K.; Thompson, J.W.; et al. Metabolic network failures in Alzheimer's disease: A biochemical road map. *Alzheimers Dement.* **2017**, *13*, 965–984. [CrossRef] [PubMed]
113. Varma, V.R.; Oommen, A.M.; Varma, S.; Casanova, R.; An, Y.; Andrews, R.M.; O'Brien, R.; Pletnikova, O.; Troncoso, J.C.; Toledo, J.; et al. Brain and blood metabolite signatures of pathology and progression in Alzheimer disease: A targeted metabolomics study. *PLoS Med.* **2018**, *15*, e1002482. [CrossRef]
114. Wilkins, J.M.; Trushina, E. Application of Metabolomics in Alzheimer's Disease. *Front. Neurol.* **2017**, *8*, 719. [CrossRef]
115. Yang, J.; Li, S.; He, X.B.; Cheng, C.; Le, W. Induced pluripotent stem cells in Alzheimer's disease: Applications for disease modeling and cell-replacement therapy. *Mol. Neurodegener.* **2016**, *11*, 39. [CrossRef] [PubMed]
116. Ryan, K.J.; White, C.C.; Patel, K.; Xu, J.; Olah, M.; Replogle, J.M.; Frangieh, M.; Cimpean, M.; Winn, P.; McHenry, A.; et al. A human microglia-like cellular model for assessing the effects of neurodegenerative disease gene variants. *Sci. Transl. Med.* **2017**, *9*, eaai7635. [CrossRef]
117. Karch, C.M.; Hernandez, D.; Wang, J.C.; Marsh, J.; Hewitt, A.W.; Hsu, S.; Norton, J.; Levitch, D.; Donahue, T.; Sigurdson, W.; et al. Human fibroblast and stem cell resource from the Dominantly Inherited Alzheimer Network. *Alzheimers Res. Ther.* **2018**, *10*, 69. [CrossRef]
118. Tcw, J. Human iPSC application in Alzheimer's disease and Tau-related neurodegenerative diseases. *Neurosci. Lett.* **2019**, *699*, 31–40. [CrossRef]
119. Xiang, X.; Piers, T.M.; Wefers, B.; Zhu, K.; Mallach, A.; Brunner, B.; Kleinberger, G.; Song, W.; Colonna, M.; Herms, J.; et al. The Trem2 R47H Alzheimer's risk variant impairs splicing and reduces Trem2 mRNA and protein in mice but not in humans. *Mol. Neurodegener.* **2018**, *13*, 49. [CrossRef] [PubMed]
120. Park, J.; Wetzel, I.; Marriott, I.; Dreau, D.; D'Avanzo, C.; Kim, D.Y.; Tanzi, R.E.; Cho, H. A 3D human triculture system modeling neurodegeneration and neuroinflammation in Alzheimer's disease. *Nat. Neurosci.* **2018**, *21*, 941–951. [CrossRef]
121. Matsui, T.K.; Tsuru, Y.; Hasegawa, K.; Kuwako, K.I. Vascularization of human brain organoids. *Stem Cells* **2021**, *39*, 1017–1024. [PubMed]
122. Morris, J.C. The Clinical Dementia Rating (CDR): Current version and scoring rules. *Neurology* **1993**, *43*, 2412–2414. [CrossRef] [PubMed]
123. Folstein, M.F.; Folstein, S.E.; Mchugh, P.R. Mini-Mental State—Practical Method for Grading Cognitive State of Patients for Clinician. *J. Psychiatr. Res.* **1975**, *12*, 189–198. [CrossRef]
124. Nasreddine, Z.S.; Phillips, N.A.; Bedirian, V.; Charbonneau, S.; Whitehead, V.; Collin, I.; Cummings, J.L.; Chertkow, H. The Montreal Cognitive Assessment, MoCA: A brief screening tool for mild cognitive impairment. *J. Am. Geriatr. Soc.* **2005**, *53*, 695–699. [CrossRef] [PubMed]
125. Rossetti, H.C.; Munro Cullum, C.; Hynan, L.S.; Lacritz, L.H. The CERAD Neuropsychologic Battery Total Score and the progression of Alzheimer disease. *Alzheimer Dis. Assoc. Disord.* **2010**, *24*, 138–142. [CrossRef] [PubMed]
126. Jack, C.R., Jr.; Knopman, D.S.; Jagust, W.J.; Shaw, L.M.; Aisen, P.S.; Weiner, M.W.; Petersen, R.C.; Trojanowski, J.Q. Hypothetical model of dynamic biomarkers of the Alzheimer's pathological cascade. *Lancet Neurol.* **2010**, *9*, 119–128. [CrossRef]
127. Cruchaga, C.; Del-Aguila, J.L.; Saef, B.; Black, K.; Fernandez, M.V.; Budde, J.; Ibanez, L.; Deming, Y.; Kapoor, M.; Tosto, G.; et al. Polygenic risk score of sporadic late-onset Alzheimer's disease reveals a shared architecture with the familial and early-onset forms. *Alzheimers Dement.* **2018**, *14*, 205–214. [CrossRef]
128. Darst, B.F.; Koscik, R.L.; Racine, A.M.; Oh, J.M.; Krause, R.A.; Carlsson, C.M.; Zetterberg, H.; Blennow, K.; Christian, B.T.; Bendlin, B.B.; et al. Pathway-Specific Polygenic Risk Scores as Predictors of Amyloid-β Deposition and Cognitive Function in a Sample at Increased Risk for Alzheimer's Disease. *J. Alzheimers Dis.* **2017**, *55*, 473–484. [CrossRef]

129. Kauppi, K.; Fan, C.C.; McEvoy, L.K.; Holland, D.; Tan, C.H.; Chen, C.H.; Andreassen, O.A.; Desikan, R.S.; Dale, A.M. Combining Polygenic Hazard Score with Volumetric MRI and Cognitive Measures Improves Prediction of Progression From Mild Cognitive Impairment to Alzheimer's Disease. *Front. Neurosci.* **2018**, *12*, 260. [CrossRef]
130. Tan, C.H.; Fan, C.C.; Mormino, E.C.; Sugrue, L.P.; Broce, I.J.; Hess, C.P.; Dillon, W.P.; Bonham, L.W.; Yokoyama, J.S.; Karch, C.M.; et al. Polygenic hazard score: An enrichment marker for Alzheimer's associated amyloid and tau deposition. *Acta Neuropathol.* **2018**, *135*, 85–93. [CrossRef]
131. Ferrari, R.; Wang, Y.; Vandrovcova, J.; Guelfi, S.; Witeolar, A.; Karch, C.M.; Schork, A.J.; Fan, C.C.; Brewer, J.B.; International, F.T.D.G.C.; et al. Genetic architecture of sporadic frontotemporal dementia and overlap with Alzheimer's and Parkinson's diseases. *J. Neurol. Neurosurg. Psychiatry* **2017**, *88*, 152–164. [CrossRef]
132. Suarez-Calvet, M.; Araque Caballero, M.A.; Kleinberger, G.; Bateman, R.J.; Fagan, A.M.; Morris, J.C.; Levin, J.; Danek, A.; Ewers, M.; Haass, C.; et al. Early changes in CSF sTREM2 in dominantly inherited Alzheimer's disease occur after amyloid deposition and neuronal injury. *Sci. Transl. Med.* **2016**, *8*, 369ra178. [CrossRef]
133. Preische, O.; Schultz, S.A.; Apel, A.; Kuhle, J.; Kaeser, S.A.; Barro, C.; Graber, S.; Kuder-Buletta, E.; LaFougere, C.; Laske, C.; et al. Serum neurofilament dynamics predicts neurodegeneration and clinical progression in presymptomatic Alzheimer's disease. *Nat. Med.* **2019**, *25*, 277–283. [CrossRef] [PubMed]
134. Nation, D.A.; Sweeney, M.D.; Montagne, A.; Sagare, A.P.; D'Orazio, L.M.; Pachicano, M.; Sepehrband, F.; Nelson, A.R.; Buennagel, D.P.; Harrington, M.G.; et al. Blood-brain barrier breakdown is an early biomarker of human cognitive dysfunction. *Nat. Med.* **2019**, *25*, 270–276. [CrossRef]
135. Escott-Price, V.; Sims, R.; Bannister, C.; Harold, D.; Vronskaya, M.; Majounie, E.; Badarinarayan, N.; GERAD/PERADES; IGAP Consortia; Morgan, K.; et al. Common polygenic variation enhances risk prediction for Alzheimer's disease. *Brain* **2015**, *138*, 3673–3684. [CrossRef] [PubMed]
136. Escott-Price, V.; Shoai, M.; Pither, R.; Williams, J.; Hardy, J. Polygenic score prediction captures nearly all common genetic risk for Alzheimer's disease. *Neurobiol. Aging* **2017**, *49*, 214.e7–214.e11. [CrossRef]
137. Jack, C.R., Jr.; Wiste, H.J.; Therneau, T.M.; Weigand, S.D.; Knopman, D.S.; Mielke, M.M.; Lowe, V.J.; Vemuri, P.; Machulda, M.M.; Schwarz, C.G.; et al. Associations of Amyloid, Tau, and Neurodegeneration Biomarker Profiles with Rates of Memory Decline Among Individuals Without Dementia. *JAMA* **2019**, *321*, 2316–2325. [CrossRef]
138. Kester, M.I.; Teunissen, C.E.; Sutphen, C.; Herries, E.M.; Ladenson, J.H.; Xiong, C.; Scheltens, P.; van der Flier, W.M.; Morris, J.C.; Holtzman, D.M.; et al. Cerebrospinal fluid VILIP-1 and YKL-40, candidate biomarkers to diagnose, predict and monitor Alzheimer's disease in a memory clinic cohort. *Alzheimers Res. Ther.* **2015**, *7*, 59. [CrossRef]
139. Casaletto, K.B.; Elahi, F.M.; Bettcher, B.M.; Neuhaus, J.; Bendlin, B.B.; Asthana, S.; Johnson, S.C.; Yaffe, K.; Carlsson, C.; Blennow, K.; et al. Neurogranin, a synaptic protein, is associated with memory independent of Alzheimer biomarkers. *Neurology* **2017**, *89*, 1782–1788. [CrossRef]
140. Nilselid, A.M.; Davidsson, P.; Nagga, K.; Andreasen, N.; Fredman, P.; Blennow, K. Clusterin; Clusterin in cerebrospinal fluid: Analysis of carbohydrates and quantification of native and glycosylated forms. *Neurochem. Int.* **2006**, *48*, 718–728. [CrossRef]
141. Piccio, L.; Buonsanti, C.; Cella, M.; Tassi, I.; Schmidt, R.E.; Fenoglio, C.; Rinker, J., 2nd; Naismith, R.T.; Panina-Bordignon, P.; Passini, N.; et al. Identification of soluble TREM-2 in the cerebrospinal fluid and its association with multiple sclerosis and CNS inflammation. *Brain* **2008**, *131*, 3081–3091. [CrossRef]
142. Heslegrave, A.; Heywood, W.; Paterson, R.; Magdalinou, N.; Svensson, J.; Johansson, P.; Ohrfelt, A.; Blennow, K.; Hardy, J.; Schott, J.; et al. Increased cerebrospinal fluid soluble TREM2 concentration in Alzheimer's disease. *Mol. Neurodegener.* **2016**, *11*, 3. [CrossRef]
143. Piccio, L.; Deming, Y.; Del-Aguila, J.L.; Ghezzi, L.; Holtzman, D.M.; Fagan, A.M.; Fenoglio, C.; Galimberti, D.; Borroni, B.; Cruchaga, C. Cerebrospinal; Cerebrospinal fluid soluble TREM2 is higher in Alzheimer disease and associated with mutation status. *Acta Neuropathol.* **2016**, *131*, 925–933. [CrossRef] [PubMed]
144. Schlepckow, K.; Kleinberger, G.; Fukumori, A.; Feederle, R.; Lichtenthaler, S.F.; Steiner, H.; Haass, C. An Alzheimer-associated TREM2 variant occurs at the ADAM cleavage site and affects shedding and phagocytic function. *EMBO Mol. Med.* **2017**, *9*, 1356–1365. [CrossRef] [PubMed]
145. Del-Aguila, J.L.; Benitez, B.A.; Li, Z.; Dube, U.; Mihindukulasuriya, K.A.; Budde, J.P.; Farias, F.H.G.; Fernandez, M.V.; Ibanez, L.; Jiang, S.; et al. TREM2 brain transcript-specific studies in AD and TREM2 mutation carriers. *Mol. Neurodegener.* **2019**, *14*, 18. [CrossRef]
146. Suarez-Calvet, M.; Capell, A.; Araque Caballero, M.A.; Morenas-Rodriguez, E.; Fellerer, K.; Franzmeier, N.; Kleinberger, G.; Eren, E.; Deming, Y.; Piccio, L.; et al. CSF progranulin increases in the course of Alzheimer's disease and is associated with sTREM2, neurodegeneration and cognitive decline. *EMBO Mol. Med.* **2018**, *10*, e9712. [CrossRef] [PubMed]
147. Toden, S.; Zhuang, J.; Acosta, A.D.; Karns, A.P.; Salathia, N.S.; Brewer, J.B.; Wilcock, D.M.; Aballi, J.; Nerenberg, M.; Quake, S.R.; et al. Noninvasive characterization of Alzheimer's disease by circulating, cell-free messenger RNA next-generation sequencing. *Sci. Adv.* **2020**, *6*, eabb1654. [CrossRef]
148. Koh, W.; Pan, W.; Gawad, C.; Fan, H.C.; Kerchner, G.A.; Wyss-Coray, T.; Blumenfeld, Y.J.; El-Sayed, Y.Y.; Quake, S.R. Noninvasive in vivo monitoring of tissue-specific global gene expression in humans. *Proc. Natl. Acad. Sci. USA* **2014**, *111*, 7361–7366. [CrossRef]

149. Sheinerman, K.S.; Toledo, J.B.; Tsivinsky, V.G.; Irwin, D.; Grossman, M.; Weintraub, D.; Hurtig, H.I.; Chen-Plotkin, A.; Wolk, D.A.; McCluskey, L.F.; et al. Circulating brain-enriched microRNAs as novel biomarkers for detection and differentiation of neurodegenerative diseases. *Alzheimers Res. Ther.* **2017**, *9*, 89. [CrossRef] [PubMed]
150. Schor, N.F. What the halted phase III γ-secretase inhibitor trial may (or may not) be telling us. *Ann. Neurol.* **2011**, *69*, 237–239. [CrossRef]
151. Szaruga, M.; Munteanu, B.; Lismont, S.; Veugelen, S.; Horre, K.; Mercken, M.; Saido, T.C.; Ryan, N.S.; De Vos, T.; Savvides, S.N.; et al. Alzheimer's-Causing Mutations Shift Abeta Length by Destabilizing γ-Secretase-Abetan Interactions. *Cell* **2017**, *170*, 443–456.e14. [CrossRef]
152. Bateman, R.J.; Benzinger, T.L.; Berry, S.; Clifford, D.B.; Duggan, C.; Fagan, A.M.; Fanning, K.; Farlow, M.R.; Hassenstab, J.; McDade, E.M.; et al. The DIAN-TU Next Generation Alzheimer's prevention trial: Adaptive design and disease progression model. *Alzheimers Dement.* **2017**, *13*, 8–19. [CrossRef]
153. Tariot, P.N.; Lopera, F.; Langbaum, J.B.; Thomas, R.G.; Hendrix, S.; Schneider, L.S.; Rios-Romenets, S.; Giraldo, M.; Acosta, N.; Tobon, C.; et al. The Alzheimer's Prevention Initiative Autosomal-Dominant Alzheimer's Disease Trial: A study of crenezumab versus placebo in preclinical PSEN1 E280A mutation carriers to evaluate efficacy and safety in the treatment of autosomal-dominant Alzheimer's disease, including a placebo-treated noncarrier cohort. *Alzheimers Dement.* **2018**, *4*, 150–160.
154. Reiman, E.M.; Langbaum, J.B.; Fleisher, A.S.; Caselli, R.J.; Chen, K.; Ayutyanont, N.; Quiroz, Y.T.; Kosik, K.S.; Lopera, F.; Tariot, P.N. Alzheimer's Prevention Initiative: A plan to accelerate the evaluation of presymptomatic treatments. *J. Alzheimers Dis.* **2011**, *26* (Suppl. S3), 321–329. [CrossRef]
155. Egan, M.F.; Kost, J.; Voss, T.; Mukai, Y.; Aisen, P.S.; Cummings, J.L.; Tariot, P.N.; Vellas, B.; van Dyck, C.H.; Boada, M.; et al. Randomized Trial of Verubecestat for Prodromal Alzheimer's Disease. *N. Engl. J. Med.* **2019**, *380*, 1408–1420. [CrossRef] [PubMed]
156. Sevigny, J.; Chiao, P.; Bussiere, T.; Weinreb, P.H.; Williams, L.; Maier, M.; Dunstan, R.; Salloway, S.; Chen, T.; Ling, Y.; et al. The antibody aducanumab reduces Abeta plaques in Alzheimer's disease. *Nature* **2016**, *537*, 50–56. [CrossRef]
157. Morris, G.P.; Clark, I.A.; Vissel, B. Inconsistencies and controversies surrounding the amyloid hypothesis of Alzheimer's disease. *Acta Neuropathol. Commun.* **2014**, *2*, 135. [CrossRef]
158. Congdon, E.E.; Sigurdsson, E.M. Tau-targeting therapies for Alzheimer disease. *Nat. Rev. Neurol.* **2018**, *14*, 399–415. [CrossRef] [PubMed]
159. Ginsberg, S.D.; Che, S.; Counts, S.E.; Mufson, E.J. Shift in the ratio of three-repeat tau and four-repeat tau mRNAs in individual cholinergic basal forebrain neurons in mild cognitive impairment and Alzheimer's disease. *J. Neurochem.* **2006**, *96*, 1401–1408. [CrossRef] [PubMed]
160. Conrad, C.; Zhu, J.; Conrad, C.; Schoenfeld, D.; Fang, Z.; Ingelsson, M.; Stamm, S.; Church, G.; Hyman, B.T. Single molecule profiling of tau gene expression in Alzheimer's disease. *J. Neurochem.* **2007**, *103*, 1228–1236. [CrossRef] [PubMed]
161. Xu, H.; Rosler, T.W.; Carlsson, T.; de Andrade, A.; Fiala, O.; Hollerhage, M.; Oertel, W.H.; Goedert, M.; Aigner, A.; Hoglinger, G.U. Tau silencing by siRNA in the P301S mouse model of tauopathy. *Curr. Gene Ther.* **2014**, *14*, 343–351. [CrossRef] [PubMed]
162. DeVos, S.L.; Miller, R.L.; Schoch, K.M.; Holmes, B.B.; Kebodeaux, C.S.; Wegener, A.J.; Chen, G.; Shen, T.; Tran, H.; Nichols, B.; et al. Tau reduction prevents neuronal loss and reverses pathological tau deposition and seeding in mice with tauopathy. *Sci. Transl. Med.* **2017**, *9*, eaag0481. [CrossRef] [PubMed]
163. Duffy, A.G.; Makarova-Rusher, O.V.; Ulahannan, S.V.; Rahma, O.E.; Fioravanti, S.; Walker, M.; Abdullah, S.; Raffeld, M.; Anderson, V.; Abi-Jaoudeh, N.; et al. Modulation of tumor eIF4E by antisense inhibition: A phase I/II translational clinical trial of ISIS 183750-an antisense oligonucleotide against eIF4E-in combination with irinotecan in solid tumors and irinotecan-refractory colorectal cancer. *Int. J. Cancer* **2016**, *139*, 1648–1657. [CrossRef]
164. Finkel, R.S.; Mercuri, E.; Darras, B.T.; Connolly, A.M.; Kuntz, N.L.; Kirschner, J.; Chiriboga, C.A.; Saito, K.; Servais, L.; Tizzano, E.; et al. Nusinersen versus Sham Control in Infantile-Onset Spinal Muscular Atrophy. *N. Engl. J. Med.* **2017**, *377*, 1723–1732. [CrossRef]
165. Yamazaki, Y.; Painter, M.M.; Bu, G.; Kanekiyo, T. Apolipoprotein E as a Therapeutic Target in Alzheimer's Disease: A Review of Basic Research and Clinical Evidence. *CNS Drugs* **2016**, *30*, 773–789. [CrossRef]
166. Cramer, P.E.; Cirrito, J.R.; Wesson, D.W.; Lee, C.Y.; Karlo, J.C.; Zinn, A.E.; Casali, B.T.; Restivo, J.L.; Goebel, W.D.; James, M.J.; et al. ApoE-directed therapeutics rapidly clear β-amyloid and reverse deficits in AD mouse models. *Science* **2012**, *335*, 1503–1506. [CrossRef]
167. Luz, I.; Liraz, O.; Michaelson, D.M. An Anti-apoE4 Specific Monoclonal Antibody Counteracts the Pathological Effects of apoE4 In Vivo. *Curr. Alzheimer Res.* **2016**, *13*, 918–929. [CrossRef]
168. Liao, F.; Li, A.; Xiong, M.; Bien-Ly, N.; Jiang, H.; Zhang, Y.; Finn, M.B.; Hoyle, R.; Keyser, J.; Lefton, K.B.; et al. Targeting of nonlipidated, aggregated apoE with antibodies inhibits amyloid accumulation. *J. Clin. Investig.* **2018**, *128*, 2144–2155. [CrossRef] [PubMed]
169. Mungenast, A.E.; Siegert, S.; Tsai, L.H. Modeling Alzheimer's disease with human induced pluripotent stem (iPS) cells. *Mol. Cell. Neurosci.* **2016**, *73*, 13–31. [CrossRef]
170. Komor, A.C.; Kim, Y.B.; Packer, M.S.; Zuris, J.A.; Liu, D.R. Programmable editing of a target base in genomic DNA without double-stranded DNA cleavage. *Nature* **2016**, *533*, 420–424. [CrossRef]

171. Pan, Y.; Short, J.L.; Choy, K.H.; Zeng, A.X.; Marriott, P.J.; Owada, Y.; Scanlon, M.J.; Porter, C.J.; Nicolazzo, J.A. Fatty Acid-Binding Protein 5 at the Blood-Brain Barrier Regulates Endogenous Brain Docosahexaenoic Acid Levels and Cognitive Function. *J. Neurosci.* **2016**, *36*, 11755–11767. [CrossRef] [PubMed]
172. Zhang, P.; Kishimoto, Y.; Grammatikakis, I.; Gottimukkala, K.; Cutler, R.G.; Zhang, S.; Abdelmohsen, K.; Bohr, V.A.; Misra Sen, J.; Gorospe, M.; et al. Senolytic therapy alleviates Abeta-associated oligodendrocyte progenitor cell senescence and cognitive deficits in an Alzheimer's disease model. *Nat. Neurosci.* **2019**, *22*, 719–728. [CrossRef] [PubMed]
173. Deshpande, P.; Gogia, N.; Singh, A. Exploring the efficacy of natural products in alleviating Alzheimer's disease. *Neural Regen. Res.* **2019**, *14*, 1321–1329. [PubMed]
174. Sarkar, A.; Irwin, M.; Singh, A.; Riccetti, M.; Singh, A. Alzheimer's disease: The silver tsunami of the 21(st) century. *Neural Regen. Res.* **2016**, *11*, 693–697. [PubMed]

Article

Pairwise Correlation Analysis of the Alzheimer's Disease Neuroimaging Initiative (ADNI) Dataset Reveals Significant Feature Correlation

Erik D. Huckvale [1], Matthew W. Hodgman [1], Brianna B. Greenwood [2], Devorah O. Stucki [2], Katrisa M. Ward [2], Mark T. W. Ebbert [1], John S. K. Kauwe [2], The Alzheimer's Disease Neuroimaging Initiative [†], The Alzheimer's Disease Metabolomics Consortium [‡] and Justin B. Miller [1,*]

[1] Sanders-Brown Center on Aging, University of Kentucky, Lexington, KY 40536, USA; Erik.Huckvale@uky.edu (E.D.H.); Matthew.Hodgman@uky.edu (M.W.H.); mark.ebbert@uky.edu (M.T.W.E.)

[2] Department of Biology, Brigham Young University, Provo, UT 84602, USA; briannabellemathre@gmail.com (B.B.G.); devorah.stucki@gmail.com (D.O.S.); katrisa14@gmail.com (K.M.W.); kauwe@byu.edu (J.S.K.K.)

* Correspondence: justin.miller@uky.edu; Tel.: +859-562-0333

[†] Data used in preparation of this article were obtained from the Alzheimer's Disease Neuroimaging Initiative (ADNI) database (adni.loni.usc.edu). As such, the investigators within the ADNI contributed to the design and implementation of ADNI and/or provided data but did not participate in analysis or writing of this report. A complete listing of ADNI investigators can be found at: http://adni.loni.usc.edu/wp-content/uploads/how_to_apply/ADNI_Acknowledgement_List.pdf (accessed on 26 September 2021).

[‡] Data used in preparation of this article were generated by the Alzheimer's Disease Metabolomics Consortium (ADMC). As such, the investigators within the ADMC provided data but did not participate in analysis or writing of this report. A complete listing of ADMC investigators can be found at: https://sites.duke.edu/adnimetab/team/ (accessed on 26 September 2021).

Citation: Huckvale, E.D.; Hodgman, M.W.; Greenwood, B.B.; Stucki, D.O.; Ward, K.M.; Ebbert, M.T.W.; Kauwe, J.S.K.; The Alzheimer's Disease Neuroimaging Initiative; The Alzheimer's Disease Metabolomics Consortium; Miller, J.B. Pairwise Correlation Analysis of the Alzheimer's Disease Neuroimaging Initiative (ADNI) Dataset Reveals Significant Feature Correlation. *Genes* **2021**, *12*, 1661. https://doi.org/10.3390/genes12111661

Academic Editor: Diego Centonze

Received: 27 September 2021
Accepted: 20 October 2021
Published: 21 October 2021

Publisher's Note: MDPI stays neutral with regard to jurisdictional claims in published maps and institutional affiliations.

Copyright: © 2021 by the authors. Licensee MDPI, Basel, Switzerland. This article is an open access article distributed under the terms and conditions of the Creative Commons Attribution (CC BY) license (https://creativecommons.org/licenses/by/4.0/).

Abstract: The Alzheimer's Disease Neuroimaging Initiative (ADNI) contains extensive patient measurements (e.g., magnetic resonance imaging [MRI], biometrics, RNA expression, etc.) from Alzheimer's disease (AD) cases and controls that have recently been used by machine learning algorithms to evaluate AD onset and progression. While using a variety of biomarkers is essential to AD research, highly correlated input features can significantly decrease machine learning model generalizability and performance. Additionally, redundant features unnecessarily increase computational time and resources necessary to train predictive models. Therefore, we used 49,288 biomarkers and 793,600 extracted MRI features to assess feature correlation within the ADNI dataset to determine the extent to which this issue might impact large scale analyses using these data. We found that 93.457% of biomarkers, 92.549% of the gene expression values, and 100% of MRI features were strongly correlated with at least one other feature in ADNI based on our Bonferroni corrected α (p-value $\leq 1.40754 \times 10^{-13}$). We provide a comprehensive mapping of all ADNI biomarkers to highly correlated features within the dataset. Additionally, we show that significant correlation within the ADNI dataset should be resolved before performing bulk data analyses, and we provide recommendations to address these issues. We anticipate that these recommendations and resources will help guide researchers utilizing the ADNI dataset to increase model performance and reduce the cost and complexity of their analyses.

Keywords: ADNI; pairwise feature correlation; feature reduction; machine learning; Alzheimer's disease

1. Introduction

Researchers increasingly leverage big data techniques, such as machine learning, to identify patterns indicative of disease trajectory to better understand, diagnose, and treat Alzheimer's disease (AD). This search for a cure has led to ever-expanding datasets that have increased in both size and complexity [1]. Although AD is a progressive neurodegenerative disorder characterized by the "A/T/N" system (i.e., β-amyloid biomarker buildup, tau biomarker buildup, and neurodegeneration or neuronal injury) [2], heterogeneity in disease manifestation and trajectory impact our ability to accurately diagnose or treat AD [3,4]. However, since AD is the most common cause of dementia [5], and related AD health-care costs are projected to exceed \$1 trillion by 2050 [6], it is imperative to leverage large biobanks to best define its etiology and search for a cure. Here, we utilize the Alzheimer's Disease Neuroimaging Initiative (ADNI) dataset, which contains patient data for AD cases and controls spanning 49,288 biomarkers and 1.2 terabytes of neuroimages.

While large biological datasets, such as the ADNI cohort, are crucial for developing accurate models, an excessive number of features can cause algorithms to take more time to compute [7–9], require significantly more computational resources [9–11], increase model complexity [9], reduce model performance [12], and ultimately increase the costs of large-scale analyses. These issues often make these types of analyses intractable for smaller research labs with limited computational resources. Researchers typically sidestep the issue by reducing their analyses to a pre-selected subset of features based on literature searches or specific hypotheses, which limits the creative exploration of other features included in the dataset. Programmatic solutions to feature selection also exist [13] but require a pairwise correlation analysis to identify redundancy [14]. Pairwise correlation analyses iteratively calculate the correlation between each feature and all other feature in the dataset [15]. When multiple features are highly correlated with each other, one feature can be used as representative of all other features, which effectively reduces the size of the dataset for downstream analyses.

We assessed correlation within the ADNI dataset to determine the extent to which machine learning might be impacted by correlated features. We performed a pairwise correlation analysis of all 49,288 biomarkers and 793,600 extracted magnetic resonance imaging (MRI) features (842,888 total features). We repeated the pairwise correlation analysis using subsets stratified by sex and clinical dementia rating (CDR) to determine if the correlated features should be interpreted broadly (i.e., across the dataset) or more narrowly (e.g., only in females). We identified high feature redundancy that impacts 99.566% of all features, including 93.457% of the ADNIMERGE features and 92.549% of the gene expression features. Additionally, we identified metadata in the ADNI tables that were not programmatically distinguishable from biomarkers, and several duplicate features with different column headers.

We propose that machine learning on the ADNI dataset should remove highly correlated or duplicate features and metadata to increase model performance, decrease model training time, and accelerate AD research toward improved understanding, diagnosis, and treatment. We provide correlation tables to facilitate the identification and filtering of highly correlated features within ADNI.

2. Materials and Methods

Data used in the preparation of this article were obtained from the Alzheimer's Disease Neuroimaging Initiative (ADNI) database (adni.loni.usc.edu) on 15 November 2019. The ADNI was launched in 2003 as a public-private partnership, led by principal investigator Michael W. Weiner, MD. The primary goal of ADNI has been to test whether serial magnetic resonance imaging (MRI), positron emission tomography (PET), other biological markers, and clinical and neuropsychological assessment can be combined to measure the progression of mild cognitive impairment (MCI) and early Alzheimer's disease (AD). For up-to-date information, see www.adni-info.org. ADNI researchers collect, validate, and utilize data, including MRI and PET images, genetics, cognitive tests, CSF and blood biomarkers as predictors of the disease.

We divided the ADNI data into three domains: the ADNIMERGE domain, which contains features such as cerebral spinal fluid (CSF) biomarkers and cognitive function test scores; the gene expression domain, which contains gene expression levels from blood microarrays [16]; and the MRI domain, which contains features we extracted from MRIs using deep convolutional autoencoders. Step-by-step protocols for each domain are included at https://github.com/jmillerlab/ADNI_Correlation and described below.

2.1. ADNIMERGE Domain

We constructed the ADNIMERGE domain using the R package, ADNIMERGE [17]. We retrieved the data from ADNIMERGE because it contains the ADNI tabular data in the form of multiple individual tables conveniently stored within a single package. To efficiently merge these data, we developed a custom method for combining the ADNIMERGE tables into one table by joining each table by its patient ID and most recent measurement (see Figure S1).

We preprocessed all tables in the ADNIMERGE domain before combining them. We capitalized all headers to have consistent feature names across tables with overlapping features. Columns with only one unique value were removed because features without variation are uninformative in machine learning. Every feature table contained patient IDs that we used as primary keys when combining tables. We removed the 'Data Dictionary' table from the domain because it contained only meta-data and no patient IDs. We recorded the data type of each feature, whether nominal or numeric, to determine which statistical tests to apply in downstream analysis. All features containing number values were marked as numeric unless they contained fewer than ten unique values, in which case they were considered nominal. Features containing text were marked as nominal. However, if those values contained more than 20 unique values, they were removed from the ADNIMERGE domain to eliminate features that were unique or almost unique for the individual, which might occur when the features are unique identifiers or notes written by the data recorders.

We further cleaned the data so that every feature had a single value for each individual. For features that contained longitudinal data, we selected the most recent value using its recording date. If the recording dates were not available, we arbitrarily selected one value for the person. If an individual did not have a value for a certain feature, we marked it as unknown. We removed features that either contained only unknown values or only one unique value because those features are uninformative in machine learning. The resulting table contained rows corresponding to each person, and columns corresponding to each ADNIMERGE feature.

Lastly, we resolved unknown values by either removing or imputing them. Features with fewer than 80% known values were removed to ensure accuracy. Nominal features with fewer than 20 patients in any of their categories were also removed as these features did not meet the assumptions of our statistical tests. Numeric values were then imputed using a Bayesian-ridge estimator [18] that predicts unknown values for numeric features based on known values of other features. The random number generator for this stochastic algorithm was seeded for reproducibility. A simple imputer was used for unknown nominal values, which replaced unknown values with the most frequent known category. These imputing algorithms were provided by the Scikit-learn Python package [19]. The completed ADNIMERGE data set contained 1131 features.

2.2. Gene Expression Domain

We downloaded the gene expression domain from ADNI, which contains a table of gene expression profiles from blood RNA and has previously been explored using machine learning [16]. All quality control and normalization were conducted by ADNI before its inclusion in the dataset. We transposed the table so that feature columns corresponded to the normalized gene expression levels for each patient. Any columns that contained metadata or did not contain a header were removed. The resulting gene expression domain contained a total of 48,157 genes.

2.3. MRI Domain

The MRI domain contained features we extracted from MRIs using deep convolutional autoencoders designed and trained using the PyTorch deep learning library [20]. All features in this dataset were numeric transformed pixel values. The image dataset initially consisted of 1.2 terabytes of MRIs, but we used only MRIs that belonged to the 743 patients also found in both the ADNIMERGE and gene expression domains. We organized these MRIs using the PyDicom Python package [21] so that each patient had a sequence of MRI images scanned from one side of the skull to the other. We used the med2image [22] Python package to convert the MRIs from DICOM format to PNG so that they could be used in deep learning. All images were resized to 128 by 128 pixels using the OpenCV Please confirm that the intended meaning has been retained.ython package [23]. Image pixel values were then normalized using min-max normalization [24] to optimize them for the deep learning model.

Each patient had a sequence of 124 sagittal MRI slices. Each of those two-dimensional images were compressed to a one-dimensional latent space of 6,400 extracted features. By storing images in one-dimensional arrays, the MRI domain could be tabular and therefore merged with the other two domains. We trained separate autoencoders for each of the 124 slice indices of the MRI sequences across all the patients using the Adam optimizer for artificial neural networks [25] (See Figure S2). The 124 latent vectors for each individual were concatenated for a total of $124 \times 6400 = 793,600$ extracted MRI features per person. These concatenated MRI features acted as the rows in the MRI domain (see Figure 1). We seeded all random number generators for reproducibility since the model training algorithms are stochastic in nature.

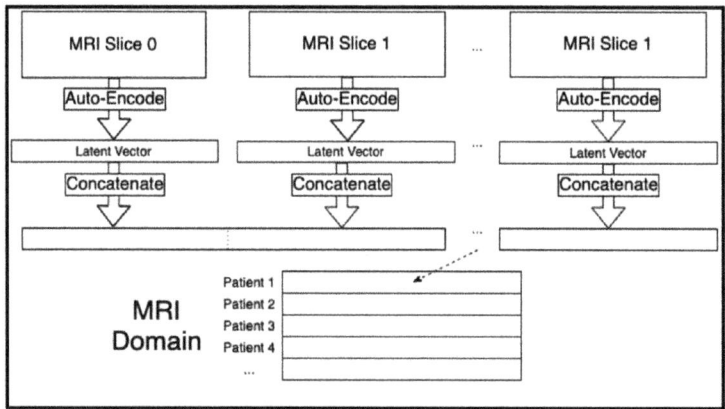

Figure 1. Creation of the MRI domain from the MRI slice sequences using the trained convolutional autoencoders. A separate autoencoder was trained for each MRI slice, and the latent space was concatenated for each person to create a row specific to that individual.

2.4. Combining All Domains

We merged the ADNIMERGE, gene expression, and MRI domains into a singular dataset that we used for our correlation analysis. The combined dataset contained a total of 1131 ADNIMERGE features + 48,157 gene features + 793,600 MRI features = 842,888 features for 743 individuals.

2.5. Correlation Analysis

We performed a pairwise correlation analysis where we compared every feature in our dataset to every other feature. For each comparison, we chose a statistical test depending on the data types of the two features as well as the normality of their distribution, if numeric (See Table 1).

Table 1. Statistical tests chosen for comparisons and their conditions.

Comparison Data Types	Condition	Statistical Test
Numeric and Numeric	Both features follow a normal distribution	Pearson correlation
Numeric and Numeric	At least one of the features does not follow a normal distribution	Spearman correlation
Categorical and Categorical	The contingency table contains at least one frequency less than five	N/A
Categorical and Categorical	All frequencies in the contingency table are greater than or equal to five	Chi-squared
Numeric and Categorical	All categories have a normal distribution	ANOVA
Numeric and Categorical	Not all categories have a normal distribution	Kruskal-Wallis

For numeric features, depending on the normality of their distribution, we chose between a parametric (normal distribution) or a non-parametric (non-normal distribution) statistical test. If both features were nominal, we used the Chi-squared test unless the contingency table resulting from the two features did not each contain at least five instances. In that case, the test was not performed.

The statistical tests and the test for normality [26] were conducted using the SciPy Python package [27]. Because we analyzed all pairwise comparisons (excluding self-comparisons) of m features across n individuals, the big-O time complexity of our algorithm was $O(n * m^2)$. However, we optimized performance by using parallel processing across four processing cores. Because of the high number of comparisons (355,229,668,828), we employed a Bonferroni corrected α value of $\alpha = 1.40754 \times 10^{-13}$ (0.05/355,229,668,828). Only feature comparisons with significant p-values were stored to save disk space.

2.6. Subset Stratification Analysis

Feature correlation within the entire dataset may occur if a subset of individuals determine that correlation. Therefore, we determined if the significantly correlated features were also correlated in different subsets stratified by sex (e.g., male or female) and clinical dementia rating (CDR; e.g., 0, 0.5, ≥1) [28]. Sex-specific AD pathologies occur [29], and pathologies vary based on cognitive status [29]. Additionally, if disease-modifying treatments in AD cases affect feature correlation, the features would not be correlated in all subsets. Therefore, if the features remain correlated in the complete dataset and each stratified subtype, they can be considered redundant.

We created five subsets from the combined dataset: female patients, male patients, cognitive normal controls where CDR = 0, patients with mild cognitive impairment where CDR = 0.5, and patients with AD where CDR ≥ 1.0. Next, we identified correlations with the highest possible significance (p-value $\leq 5 \times 10^{-324}$, which is the smallest positive value in Python 3.7) from the original analysis of the combined dataset. We reran those correlation analyses within each of the five subsets to determine if certain stratifications affected the correlation significance. We noted that significance will drop due to smaller sample size in the comparison and some features were dropped from the analysis (e.g., features with only one unique value in the subset). In practice, only the AD subset experienced loss of significant comparisons as a result of sub-setting.

3. Results

We found that 839,226 ADNI features (99.566% of the total number of features) are significantly correlated with at least one other feature (93.457% of the ADNIMERGE features, 92.549% of the gene expression features, and 100% of the MRI features). Table 2 shows a subset of features that are correlated at the highest significance threshold with more than one other feature, including patient sex, intra-cranial volume, various neuropsychological batteries, ventral diencephalon volume, and cerebrospinal fluid (CSF) glucose levels.

Table 2. ADNIMERGE Features that are Highly Correlated with other Features.

Feature Name	ADNIMERGE Frequency	Gene Expression Frequency	MRI Frequency	Total Frequency
PTGENDER	207	281	145,780	146268
ICV	265	84	143,377	143,724
CLOCKNUM	199	0	97,725	97,924
COPYTIME	243	0	97,030	97,228
CLOCKSYM	216	0	96,307	96,523
ST65SV	191	46	81,250	81,487
GLUCOSE	155	0	81,245	81,400

Numbers of correlated features for seven example features in the ADNIMERGE domain. The 'Feature Name' is the column header as it appeared in our constructed tabular data set. The 'ADNIMERGE Frequency' is the number of ADNIMERGE features that are highly correlated with the feature. For example, intra-cranial volume (ICV) is correlated with 265 other ADNIMERGE features. It is likewise correlated with 84 gene expression levels and 143,377 extracted MRI features. The 'Total Frequency' is the sum of the 'ADNIMERGE Frequency', 'Gene Expression Frequency', and 'MRI Frequency'. In other words, it is the total number of features that are highly correlated with each row across the entire ADNI data set.

While Table 2 shows the numbers of correlated features for seven example features from the ADNIMERGE domain, Table S1 shows the same but for all the ADNIMERGE features. Both Table 2 and Table S1 show the numbers of correlated features based on our Bonferroni corrected α (p-value $\leq 1.40754 \times 10^{-13}$). However, Table S2 shows the numbers of correlated features based on the maximally significant α (p-value $\leq 5 \times 10^{-324}$). The complete table (gene expression and MRI features included in addition to ADNIMERGE) for the Bonferroni corrected α is available online at: https://github.com/jmillerlab/ADNI_Correlation/blob/main/data/sig-freqs/bonferroni-sig-freqs.csv.

The complete table for the maximally significant α is available online at: https://github.com/jmillerlab/ADNI_Correlation/blob/main/data/sig-freqs/maximum-sig-freqs.csv.

While the complete tables containing our results are available online, we provide a summary of those results in Table 3 (Bonferroni corrected α) and Table 4 (maximally significant α).

Table 3. Summarized correlated feature frequencies based on the Bonferroni corrected α.

A—ADNIMERGE Frequencies					B—Gene Expression Frequencies				
Domain	Average	Standard Deviation	Minimum	Maximum	Domain	Average	Standard Deviation	Minimum	Maximum
ADNIMERGE	129.49	88.06	1	346	ADNIMERGE	11.91	30.9	0	616
Gene Expression	0.28	5.52	0	189	Gene Expression	6139.72	6195.45	1	24,588
MRI	9.31	20.09	0	188	MRI	7.87	19.66	0	149
C—MRI Frequencies					D—Total Frequencies				
Domain	Average	Standard Deviation	Minimum	Maximum	Domain	Average	Standard Deviation	Minimum	Maximum
ADNIMERGE	6988.04	19,170.23	0	145,780	ADNIMERGE	7129.43	19,203.93	1	146,268
Gene Expression	140.05	3642.09	0	119,556	Gene Expression	6280.05	7096.48	1	120,141
MRI	141,348.57	69,866.96	81	347,944	MRI	141,365.75	69,873.31	81	347,955

Summary of the numbers of correlated features based on the Bonferroni corrected α. Sections A through D provide summary statistics for the domain frequencies for ADNIMERGE, Gene Expression, MRI, and Total. For example, the meaning of the 'Average' column and 'MRI' row in table A is the average number of ADNIMERGE features with which the MRI features are strongly correlated. That row states that the MRI features are strongly correlated with an average of 9.31 ADNIMERGE features with a standard deviation of 20.9 features. The 0 in the 'Minimum' column indicates that at least one MRI feature is not correlated with any ADNIMERGE features. The 188 under 'Maximum' indicates that at least one MRI feature is correlated with 188 ADNIMERGE features when p-value $\leq 1.40754 \times 10^{-13}$.

Table 4. Summarized correlated feature frequencies based on the maximally significant α.

A—ADNIMERGE Frequencies					B—Gene Expression Frequencies				
Domain	Average	Standard Deviation	Minimum	Maximum	Domain	Average	Standard Deviation	Minimum	Maximum
ADNIMERGE	5.48	5.83	1	23	ADNIMERGE	0.0	0.0	0	0
Gene Expression	0.0	0.0	0	0	Gene Expression	1.55	0.94	1	7
MRI	0.0	0.0	0	0	MRI	0.0	0.0	0	0
C—MRI Frequencies					D—Total Frequencies				
Domain	Average	Standard Deviation	Minimum	Maximum	Domain	Average	Standard Deviation	Minimum	Maximum
ADNIMERGE	0.0	0.0	0	0	ADNIMERGE	5.48	5.83	1	23
Gene Expression	0.0	0.0	0	0	Gene Expression	1.55	0.94	1	7
MRI	2457.08	4397.99	1	11957	MRI	2457.08	4397.99	1	11,957

Summary of the numbers of correlated features based on the maximally significant comparisons (p-value $\leq 5 \times 10^{-324}$). Interestingly, when applying a maximally significant α, features were only strongly correlated with other features in their same domain.

3.1. Domain-Specific Correlation Analysis Results

First, we report the number of times ADNIMERGE features correlated with each domain-specific feature in the dataset, after correcting for multiple testing using a Bonferroni α value (Table 3A). Although many gene expression probes were not highly correlated with ADNIMERGE features (mean = 0.28 ± 5.52), gene expression for ubiquitin specific peptidase 9 Y-linked (*USP9Y*; probe set: 11725293_at) was correlated with the 189 ADNIMERGE features. In contrast, many ADNIMERGE features were highly correlated with other ADNIMERGE features (mean = 129.49 ± 88.06 highly correlated comparisons per feature). The volume (cortical parcellation) of right fusiform (ADNIMERGE header: ST85CV) was correlated with 346 ADNIMERGE features, which was the highest frequency. Similarly, MRI features displayed high correlation with many ADNIMERGE data (mean = 9.31 ± 20.09), and one MRI feature was highly correlated with 188 ADNIMERGE features.

Next, we report how many gene expression features correlate with each domain-specific feature using a Bonferroni corrected α (Table 3). For ADNIMERGE features, the patient date of birth correlates with the most Affymetrix probes (616 gene probes), and the mean number of significant comparisons per feature was 11.91 ± 30.9. The MRI dataset also contained many significant correlations (mean = 7.87 ± 19.66), and one feature from the MRI autoencoder correlated with 149 gene expression values. Gene expression features, on average, strongly correlated with 6139.72 ± 6195.45 other gene expression probes. The probe expression levels for adducin 2 (*ADD3*; probe set: 11721606_a_at) are strongly correlated with the most other gene probes (24,588 probes), which consists of 51.058% of the total number of probes in the dataset.

Finally, we report the number of significant correlations with the MRI-extracted features from the autoencoder (Table 3). A single gene expression probe for the X-inactive specific transcript (*XIST*) non-protein coding region (probe set: 11757857_s_at) was strongly correlated with 119,556 MRI features, and gene probes, on average, were highly correlated with 140.05 ± 3642.09 MRI features. The ADNIMERGE features were correlated with 6988.04 ± 19,170.23 MRI features and patient sex had the highest number of significant correlations with MRI features (145,780 significant correlations). MRI to MRI feature redundancy is even higher, with a single MRI feature being correlated with 347,944 other MRI features, and each MRI feature had significant feature redundancy (minimum = 81; mean = 141,348.57 ± 69,866.96).

Similar analyses were conducted on the total frequencies of feature redundancy (Table 3) and using the maximum significance threshold (Table 4).

3.2. Subset Stratification Analysis Results

Tables S3–S7 contain the summary statistics for the five subsets. Differences in these tables compared to Table 4 indicate that comparisons that were maximally significant using the entire data set (all 743 patients) were not maximally significant for a given subset. We found that the male, female, and AD subsets each exhibited fewer maximally significant feature comparisons than found in the complete correlation analysis (compare Tables S3, S4, and S7 with Table 4). Features lost an average of 1391.485 ± 2835.093 features that they were maximally correlated with (i.e., based on the maximally significant α of 5×10^{-324}) after correlation analysis on the stratified male subset. Likewise, features lost an average of 1323.918 ± 2694.103 maximally significant correlations after analysis on the female subset. Finally, an average of 2343.442 ± 4325.865 correlations per feature were lost using the AD subset. Conversely, for the healthy control and mild cognitive impairment subsets, all feature comparisons maintained maximum significance (compare Tables S5 and S6 to Table 4). While it is a small amount, some of the loss in the AD subset is attributed to the sub-setting itself due to features in the subset no longer having more than one unique value or no longer satisfying the assumptions of our statistical tests. Tables S8–S12 (where each table represents a different subset) show that the sub-setting alone resulted in no loss of maximally significant comparisons except in the AD subset since the AD subset table is the only one with values that differ from Table 4. Specifically, features lost an average of 2.643 ± 0.005 correlations due to the AD subset stratification alone (i.e., the small sample size resulted in fewer possible correlation comparisons).

3.3. Feature-Correlation Mappings

We created a mapping from each gene expression and ADNIMERGE feature to a list of gene expression and ADNIMERGE features with which they are strongly correlated (based on the Bonferroni corrected α). Table S13 shows the computational resources used to conduct these comparisons. There were 1,214,628,828 comparisons (0.342% of all comparisons) that involved only ADNIMERGE features or gene expression features and our mapping took up 1.87 gigabytes of disk space. We excluded MRI features from the mapping because autoencoder features do not have clear biological significance and filesharing size restraints would preclude including those comparisons online (~550 gigabytes of disk space). The ADNIMERGE and gene mappings are available as a downloadable Python pickle file. We chose the pickle file format because it facilitates easy integration with Python scripts and research pipelines. This file is available online at: https://drive.google.com/file/d/1uRuT6rhDVDeeBuRYPif3Ate3u1UVs-hO/view?usp=sharing.

4. Discussion

Our results demonstrate a high amount of feature redundancy in the ADNI dataset that should be considered when using the dataset for machine learning. While we make no claims about feature correlation in other large-scale databanks, the significant feature correlation in ADNI suggests that this issue might be more widespread than previously thought and should be considered before performing large-scale data mining. For example, a single gene expression feature alone could replace more than half the gene expression values because it is highly correlated with expression in each of those genes. Thousands of MRI features can be replaced by ADNIMERGE or gene expression features, and the MRI features themselves can be further reduced. A single MRI feature can replace up to 43.844% of the MRI domain, indicating that we could almost double the MRI compression ratio. This redundancy may inhibit the types of analyses possible in research laboratories with limited computational resources. Furthermore, laboratories using the ADNI features for large-scale data analyses are likely to waste computational time and resources if they do not properly deal with feature redundancy within the dataset. Beyond that, models analyzing redundant ADNI data are expected to perform and generalize poorly because of the curses of dimensionality [30] and overfitting [31]. To help alleviate these issues, we provided future researchers with a mapping of highly correlated features within the ADNIMERGE and gene expression domains. We recommend that researchers using the ADNIMERGE and gene expression data download our mapping file, and we inform researchers using the MRI features of their high redundancy. Future work can include further reducing the size of the MRI domain using a 1D autoencoder, as compared to the 2D autoencoder we used to perform the initial reduction. This architectural change would likely be useful because while 2D convolutional autoencoders compress 2-dimensional data (e.g., MRIs), 1D convolutional autoencoders compress 1-dimensional data (e.g., our extracted MRI features).

Another benefit of our correlation analysis is that it shows the possibility of AD-related features, which are more costly to collect, being replaced by less costly features. For example, collecting CSF biomarkers requires intrusive lumbar punctures [32], and certain cognitive tests (i.e., Mini-Mental State Examination, Alzheimer's Disease Assessment Scale-Cognitive Subscale, Frontal Assessment Battery, etc.) are time-intensive and can be stressful for patients [33]. If such non-ideal features are strongly correlated with more palatable features, they can be replaced by data that is easier to collect. Gene expression and MRI features are strongly correlated with hundreds of ADNIMERGE features. There are varying costs in obtaining these ADNIMERGE features (e.g., time, emotional toll, and money). If such biomarkers or tests were more demanding than a relatively simple blood [14] test or MRI scan, they could be replaced by other highly correlated features that are less costly. Doing so may decrease the burden on both patients and caretakers by limiting the number of tests performed or surveys taken, as well as the amount of paperwork that needs to be completed. Additionally, this knowledge may decrease the overall costs of conducting a clinical trial or establishing a cohort if a specific test is no longer required because it does not provide additional data beyond other testing.

Furthermore, our analyses revealed issues with the ADNI data that obstruct data analysis. First, there were several features we discovered to be highly correlated with others but were merely meta-data. These features include the date a measurement was recorded, the version of ADNI when the measurement took place, or identification numbers such as bar-code, sample-identification, Ionis ID, image UID etc. While these features serve an important function in the data, they do not have biological or cognitive meaning. We recommend that such features are labeled and distinguished, so that computational researchers can programmatically identify them in their scripts and separate them from the rest of the data. We recommend the same for columns in ADNI tables that appear to be notes taken by the data recorders. ADNIMERGE could include a two-column table that maps features to their designation (e.g., metadata, biomarker, MRI, etc.).

Another issue we identified was that certain features were maximally correlated with other features that had only slight deviations in their names and are likely duplicates in the dataset. For example, Table 5 shows that two features representing intracranial volume had exactly equal results but different header names.

Table 5. Example of Features with Identical Results but Slightly Different Names.

Feature	ADNIMERGE Frequency	Gene Expression Frequency	MRI Frequency	Total Frequency	Domain
ICV	265	84	141,676	142,025	ADNIMERGE
ICV.BL	265	84	141,676	142,025	ADNIMERGE

We suspect that such features are equivalent, but they have different header names when appearing in two different tables. If multiple tables contain columns with the same features, we recommend that such columns are correctly labeled by having the exact same header name across all tables in which they appear.

We recognize that the ADNI dataset contains longitudinal data that may result in different levels of feature correlation at different time points. We chose the last recorded time series datapoint for each feature to ensure that analyzed features were collected at similar points in disease progression. Since we conducted subtype analyses that show no difference in correlation between sex or cognitive decline, feature correlation can be interpreted broadly at the population level. However, we recognize that certain limitations in the ADNI dataset (e.g., age at first patient measurement, incomplete time series data, imputation, etc.) may limit our ability to detect changes in feature correlation for individuals. Additionally, deviations from feature correlation across a time series for an individual may warrant further investigation for its disease association.

The loss of significant statistical tests performed on the subsets may be partially a result of the reduced sample size. Patients with AD represent the smallest subset, therefore having the lowest statistical power, which may explain why the AD subset had the largest drop in significant tests. However, neither the healthy control subset nor the mild cognitive impairment subset experienced any loss of significant comparisons despite being smaller than the male subset, which did lose significant comparisons. This retention of comparison frequencies suggests that other factors beyond sample size contribute to the loss of statistically significant comparisons and that some feature comparisons lose significance when stratified by sex. Similarly, our results show that comparisons become less significant when stratified by CDR and performed on the AD subset, which does not occur in the control or mild cognitive impairment subsets. The reduction in features that are testable in each subset contributes slightly to this reduction in significant comparisons but does not account for all differences (see Table S12). Therefore, not all strongly correlated feature pairs remain correlated in each CDR subset.

5. Conclusions

Our analyses contribute significantly to future AD research by exploring feature correlation within the ADNI dataset. We identified many non-ic ADNI features that are highly correlated with each other and can be replaced when building large data models. We provide a template for constructing a convolutional autoencoder capable of extracting tabular features from MRIs and inform future researchers of the redundancy among these MRI features. Additionally, we propose solutions to address feature redundancy within the non-MRI features by downloading our feature redundancy tables. We validated these correlations by sub-setting the ADNI dataset and found that most highly correlated features remain highly correlated in each stratified subset. Additionally, we propose that researchers who design clinical trials or testing for AD should be mindful of feature redundancy to reduce the unnecessary testing burden on patients and caregivers when the tests do not elicit additional information. We anticipate that this research will help guide researchers using machine learning on the ADNI dataset to take into account feature redundancy in the future.

Supplementary Materials: The following are available online at https://www.mdpi.com/article/1 0.3390/genes12111661/s1, Supplementary Figures: Figure S1: Steps for Creating the Raw Data Set for the ADNIMERGE Domain; Figure S2: Diagram Displaying the Creation of the Convolutional Autoencoders. Supplementary Tables: Table S1: Significant Comparison Frequencies For Features By Bonferroni Alpha (ADNIMERGE Features Only); Table S2: Significant Comparison Frequencies For Features By Maximum Alpha (ADNIMERGE Features Only); Table S3: Comparison Frequencies of The Subset Analysis (Male Subset); Table S4: Comparison Frequencies Of The Subset Analysis (Female Subset); Table S5: Comparison Frequencies Of The Subset Analysis (CDR = 0 Subset); Table S6: Comparison Frequencies Of The Subset Analysis (CDR = 0.5 Subset); Table S7: Comparison Frequencies Of The Subset Analysis (CDR $\geq$ 1.0 Subset); Table S8: Maximally-significant correlations with sufficient data to perform subsetting: Males; Table S9: Maximally-significant correlations with sufficient data to perform subsetting: Females; Table S10: Maximally-significant correlations with sufficient data to perform subsetting: CDR = 0; Table S11: Maximally-significant correlations with sufficient data to perform subsetting: CDR = 0.5; Table S12: Maximally-significant correlations with sufficient data to perform subsetting: CDR $\geq$ 1.0; Table S13: Resource Usage.

Author Contributions: Conceptualization, J.B.M.; data curation, D.O.S., K.M.W.; formal analysis, E.D.H., B.B.G., D.O.S. and J.B.M.; funding acquisition, M.T.W.E., J.S.K.K. and J.B.M.; investigation, E.D.H., M.W.H., D.O.S., K.M.W. and J.B.M.; methodology, E.D.H., B.B.G. and J.B.M.; project administration, M.T.W.E., J.S.K.K. and J.B.M.; resources, M.T.W.E., J.S.K.K., Alzheimer's Disease Neuroimaging Initiative ADNI, ADMC and J.B.M.; software, E.D.H., B.B.G., D.O.S. and K.M.W.; supervision, M.T.W.E., J.S.K.K. and J.B.M.; validation, E.D.H., M.W.H. and J.B.M.; visualization, E.D.H.;

writing—original draft, E.D.H. and J.B.M.; writing—review & editing, E.D.H., M.W.H., B.B.G., D.O.S., K.M.W., M.T.W.E., J.S.K.K. and J.B.M. All authors have read and agreed to the published version of the manuscript.

Funding: This work was supported by the BrightFocus Foundation [A2020118F to Miller] and the National Institutes of Health [1P30AG072946-01 to the University of Kentucky Alzheimer's Disease Research Center]. Data collection and sharing for this project was funded by the National Institute on Aging [R01AG046171, RF1AG051550 and 3U01AG024904-09S4 to the Alzheimer's Disease Metabolomics Consortium].

Institutional Review Board Statement: Not applicable.

Informed Consent Statement: Not applicable.

Data Availability Statement: All custom scripts for processing and analyzing our data are available online at: https://github.com/jmillerlab/ADNI_Correlation. The tables containing the numbers of correlated features for each feature in our constructed data set are available online at: Bonferroni corrected α: https://github.com/jmillerlab/ADNI_Correlation/blob/main/data/sig-freqs/bonferroni-sig-freqs.csv. Maximally significant α: https://github.com/jmillerlab/ADNI_Correlation/blob/main/data/sig-freqs/maximum-sig-freqs.csv. The pickle file containing the mapping of non-MRI features to the non-MRI features that they are correlated with is available online at: https://drive.google.com/file/d/1uRuT6rhDVDeeBuRYPif3Ate3u1UVs-hO/view?usp=sharing.

Acknowledgments: We thank the donors to the BrightFocus Foundation for their contributions to this research. We also acknowledge the Sanders-Brown Center on Aging at the University of Kentucky, Brigham Young University, and the Office of Research Computing at Brigham Young University for their institutional support and resources. Data collection and sharing for this project was funded by the Alzheimer's Disease Neuroimaging Initiative (ADNI) (National Institutes of Health Grant U01 AG024904) and DOD ADNI (Department of Defense award number W81XWH-12-2-0012). ADNI is funded by the National Institute on Aging, the National Institute of Biomedical Imaging and Bioengineering, and through generous contributions from the following: AbbVie, Alzheimer's Association; Alzheimer's Drug Discovery Foundation; Araclon Biotech; BioClinica, Inc.; Biogen; Bristol-Myers Squibb Company; CereSpir, Inc.; Cogstate; Eisai Inc.; Elan Pharmaceuticals, Inc.; Eli Lilly and Company; EuroImmun; F. Hoffmann-La Roche Ltd. and its affiliated company Genentech, Inc.; Fujirebio; GE Healthcare; IXICO Ltd.; Janssen Alzheimer Immunotherapy Research & Development, LLC.; Johnson & Johnson Pharmaceutical Research & Development LLC.; Lumosity; Lundbeck; Merck & Co., Inc.; Meso Scale Diagnostics, LLC.; NeuroRx Research; Neurotrack Technologies; Novartis Pharmaceuticals Corporation; Pfizer Inc.; Piramal Imaging; Servier; Takeda Pharmaceutical Company; and Transition Therapeutics. The Canadian Institutes of Health Research is providing funds to support ADNI clinical sites in Canada. Private sector contributions are facilitated by the Foundation for the National Institutes of Health (www.fnih.org). The grantee organization is the Northern California Institute for Research and Education, and the study is coordinated by the Alzheimer's Therapeutic Research Institute at the University of Southern California. ADNI data are disseminated by the Laboratory for Neuro Imaging at the University of Southern California. Data used in the preparation of this article were obtained from the Alzheimer's Disease Neuroimaging Initiative (ADNI) database (adni.loni.usc.edu). The ADNI was launched in 2003 as a public-private partnership, led by Principal Investigator Michael W. Weiner, MD. The primary goal of ADNI has been to test whether serial magnetic resonance imaging (MRI), positron emission tomography (PET), other biological markers, and clinical and neuropsychological assessment can be combined to measure the progression of mild cognitive impairment (MCI) and early Alzheimer's disease (AD). For up-to-date information, see www.adni-info.org.

Conflicts of Interest: The authors declare no conflict of interest.

References

1. Zhang, R.; Simon, G.; Yu, F. Advancing Alzheimer's Research: A Review of Big Data Promises. *Int. J. Med. Inform.* **2017**, *106*, 48–56. [CrossRef]
2. Jack, C.R.; Bennett, D.A.; Blennow, K.; Carrillo, M.C.; Feldman, H.H.; Frisoni, G.B.; Hampel, H.; Jagust, W.J.; Johnson, A.; Knopman, D.S.; et al. A/T/N: An unbiased descriptive classification scheme for Alzheimer disease biomarkers. *Neurology* **2016**, *87*, 539–547. [CrossRef] [PubMed]

3. Lam, B.; Masellis, M.; Freedman, M.; Stuss, D.T.; Black, S.E. Clinical, imaging, and pathological heterogeneity of the Alzheimer's disease syndrome. *Alzheimer's Res. Ther.* **2013**, *5*, 1. [CrossRef] [PubMed]
4. Ritchie, K.; Carrière, I.; Berr, C.; Amieva, H.; Dartigues, J.F.; Ancelin, M.L.; Ritchie, C.W. The clinical picture of Alzheimer's disease in the decade before diagnosis: Clinical and biomarker trajectories. *J. Clin. Psychiatry* **2016**, *77*. [CrossRef]
5. Ang, T.F.; An, N.; Ding, H.; Devine, S.; Auerbach, S.H.; Massaro, J.; Joshi, P.; Liu, X.; Liu, Y.; Mahon, E.; et al. Using data science to diagnose and characterize heterogeneity of Alzheimer's disease. *Alzheimer's Dement. Transl. Res. Clin. Interv.* **2019**, *5*, 264–271. [CrossRef] [PubMed]
6. Fiandaca, M.S.; Mapstone, M.E.; Cheema, A.K.; Federoff, H.J. The critical need for defining preclinical biomarkers in Alzheimer's disease. *Alzheimers Dement* **2014**, *10*, S196–S212. [CrossRef]
7. Forman, G.; Zhang, B. Distributed data clustering can be efficient and exact. *ACM SIGKDD Explor. Newsl.* **2000**, *2*, 34–38. [CrossRef]
8. Hünich, D.; Müller-Pfefferkorn, R. Managing large datasets with iRODS—A performance analysis. In Proceedings of the International Multiconference on Computer Science and Information Technology, Wisla, Poland, 18–20 October 2010.
9. Ur Rehman, M.H.; Liew, C.S.; Abbas, A.; Jayaraman, P.P.; Wah, T.Y.; Khan, S.U. Big data reduction methods: A survey. *Data Sci. Eng.* **2016**, *1*, 265–284. [CrossRef]
10. Schadt, E.E.; Linderman, M.D.; Sorenson, J.; Lee, L.; Nolan, G.P. Computational solutions to large-scale data management and analysis. *Nat. Rev. Genet.* **2010**, *11*, 647–657. [CrossRef]
11. Basney, J.; Livny, M.; Mazzanti, P. Utilizing widely distributed computational resources efficiently with execution domains. *Comput. Phys. Commun.* **2001**, *140*, 246–252. [CrossRef]
12. Sharma, N.; Saroha, K. Study of dimension reduction methodologies in data mining. In Proceedings of the International Conference on Computing, Communication Automation, Greater Noida, India, 15–16 May 2015.
13. Saeys, Y.; Inza, I.; Larranaga, P. A review of feature selection techniques in bioinformatics. *Bioinformatics* **2007**, *23*, 2507–2517. [CrossRef]
14. Chen, Z.; Wu, C.; Zhang, Y.; Huang, Z.; Ran, B.; Zhong, M.; Lyu, N. Feature selection with redundancy-complementariness dispersion. *Knowl.-Based Syst.* **2015**, *89*, 203–217. [CrossRef]
15. Yu, L.; Liu, H. Feature selection for high-dimensional data: A fast correlation-based filter solution. In Proceedings of the 20th International Conference on Machine Learning (ICML-03), Washington, DC, USA, 21–24 August 2003.
16. Miller, J.B.; Kauwe, J.S. Predicting Clinical Dementia Rating Using Blood RNA Levels. *Genes* **2020**, *11*, 706. [CrossRef]
17. ADNIMERGE: Alzheimer's Disease Neuroimaging Initiative. Available online: https://adni.bitbucket.io (accessed on 26 September 2021).
18. Tipping, M.E. Sparse Bayesian learning and the relevance vector machine. *J. Mach. Learn. Res.* **2001**, *1*, 211–244.
19. Pedregosa, F.; Varoquaux, G.; Gramfort, A.; Michel, V.; Thirion, B.; Grisel, O.; Blondel, M.; Prettenhofer, P.; Weiss, R.; Dubourg, V.; et al. Scikit-learn: Machine Learning in Python. *J. Mach. Learn. Res.* **2011**, *12*, 2825–2830.
20. Paszke, A.; Gross, S.; Massa, F.; Lerer, A.; Bradbury, J.; Chanan, G.; Killeen, T.; Lin, Z.; Gimelshein, N.; Antiga, L.; et al. PyTorch: An Imperative Style, High-Performance Deep Learning Library. *Adv. Neural Inf. Process. Syst.* **2019**, *32*, 8026–8037.
21. Mason, D. SU-E-T-33: Pydicom: An open source DICOM library. *Med. Phys.* **2011**, *38*, 3493. [CrossRef]
22. Pienaar, R. 2020. Available online: https://github.com/FNNDSC/med2image (accessed on 26 September 2021).
23. Bradski, G. The OpenCV Library. *Dr. Dobb's J. Softw. Tools* **2000**, *25*, 120–123.
24. Patro, S.; Sahu, K.K. Normalization: A preprocessing stage. *arXiv* **2015**, arXiv:1503.06462. [CrossRef]
25. Kingma, D.P.; Ba, J. Adam: A method for stochastic optimization. *arXiv* **2014**, arXiv:1412.6980.
26. D'AGOSTINO, R.; Pearson, E.S. Tests for departure from normality. Empirical results for the distributions of b2 and $\sqrt{b}$. *Biometrika* **1973**, *60*, 613–622. [CrossRef]
27. Virtanen, P.; Gommers, R.; Oliphant, T.E.; Haberland, M.; Reddy, T.; Cournapeau, D.; Burovski, E.; Peterson, P.; Weckesser, W.; Bright, J.; et al. SciPy 1.0: Fundamental Algorithms for Scientific Computing in Python. *Nat. Methods* **2020**, *17*, 261–272. [CrossRef]
28. Morris, J.C. The clinical dementia rating (cdr): Current version and. *Young* **1991**, *41*, 1588–1592.
29. Besser, L.M.; Mock, C.; Teylan, M.A.; Hassenstab, J.; Kukull, W.A.; Crary, J.F. Differences in cognitive impairment in primary age-related tauopathy versus Alzheimer disease. *J. Neuropathol. Exp. Neurol.* **2019**, *78*, 219–228. [CrossRef]
30. Kuo, F.Y.; Sloan, I.H. Lifting the curse of dimensionality. *Not. AMS* **2005**, *52*, 1320–1328.
31. Liu, R.; Gillies, D.F. Overfitting in linear feature extraction for classification of high-dimensional image data. *Pattern Recognit.* **2016**, *53*, 73–86. [CrossRef]
32. Veerabhadrappa, B.; Delaby, C.; Hirtz, C.; Vialaret, J.; Alcolea, D.; Lleó, A.; Fortea, J.; Santosh, M.S.; Choubey, S.; Lehmann, S. Detection of amyloid beta peptides in body fluids for the diagnosis of alzheimer's disease: Where do we stand? *Crit. Rev. Clin. Lab. Sci.* **2020**, *57*, 99–113. [CrossRef]
33. Oyama, A.; Takeda, S.; Ito, Y.; Nakajima, T.; Takami, Y.; Takeya, Y.; Yamamoto, K.; Sugimoto, K.; Shimizu, H.; Shimamura, M.; et al. Novel method for rapid assessment of cognitive impairment using high-performance eye-tracking technology. *Sci. Rep.* **2019**, *9*, 12932. [CrossRef]

Review

Genetic Insights into the Impact of Complement in Alzheimer's Disease

Megan Torvell [1,2,†], Sarah M. Carpanini [1,2,†], Nikoleta Daskoulidou [1,2], Robert A. J. Byrne [1,2], Rebecca Sims [3] and B. Paul Morgan [1,2,*]

1. UK Dementia Research Institute Cardiff, School of Medicine, Cardiff University, Cardiff CF24 4HQ, UK; TorvellM@Cardiff.ac.uk (M.T.); CarpaniniS@Cardiff.ac.uk (S.M.C.); daskoulidoun1@cardiff.ac.uk (N.D.); ByrneR8@cardiff.ac.uk (R.A.J.B.)
2. Division of Infection and Immunity, Systems Immunity Research Institute, School of Medicine, Cardiff University, Cardiff CF14 4XN, UK
3. Division of Psychological Medicine and Clinical Neuroscience, School of Medicine, Cardiff University, Cardiff CF24 4HQ, UK; SimsRC@cardiff.ac.uk
* Correspondence: Morganbp@cardiff.ac.uk
† These authors contributed equally to this work.

Abstract: The presence of complement activation products at sites of pathology in post-mortem Alzheimer's disease (AD) brains is well known. Recent evidence from genome-wide association studies (GWAS), combined with the demonstration that complement activation is pivotal in synapse loss in AD, strongly implicates complement in disease aetiology. Genetic variations in complement genes are widespread. While most variants individually have only minor effects on complement homeostasis, the combined effects of variants in multiple complement genes, referred to as the "complotype", can have major effects. In some diseases, the complotype highlights specific parts of the complement pathway involved in disease, thereby pointing towards a mechanism; however, this is not the case with AD. Here we review the complement GWAS hits; *CR1* encoding complement receptor 1 (CR1), *CLU* encoding clusterin, and a suggestive association of *C1S* encoding the enzyme C1s, and discuss difficulties in attributing the AD association in these genes to complement function. A better understanding of complement genetics in AD might facilitate predictive genetic screening tests and enable the development of simple diagnostic tools and guide the future use of anti-complement drugs, of which several are currently in development for central nervous system disorders.

Keywords: complement; complement receptor 1; clusterin; late-onset Alzheimer's disease; genetics; neuroinflammation

1. Alzheimer's Disease, Inflammation, and Complement

Alzheimer's disease (AD) is a common, chronic neurodegenerative disease. There are currently over 50 million cases of AD worldwide, and with an increasingly ageing population, this number will increase further [1]. AD is associated with a progressive decline in cognitive function and memory and a reduced ability to carry out day-to-day tasks, culminating in a complete loss of independence. Pathologically, AD is characterised by a build-up of protein deposits (amyloid-β (Aβ) plaques and hyperphosphorylated tau tangles) throughout the brain. Cognitive impairment is a consequence of regional neuronal and synapse loss. These events are accompanied by an inflammatory response: astrocytes and microglia, the innate immune cells of the brain, adopt a neurotoxic, phagocytic, proinflammatory phenotype and interact with plaques, tangles, and damaged or dying neurons [2]. It is increasingly apparent that the neuroinflammatory response is a driving force in AD pathology rather than a bystander or consequence of disease; perhaps the clearest evidence comes from genetic studies. Many of the genes most strongly associated with AD risk are involved in inflammation and immunity [3,4]. These data make it imperative to better understand when, where and how inflammation occurs in the course of AD in order to design better tests and novel drugs.

The complement system is an important component of the innate immune system and a potent driver of inflammation; it is the first line of defence against invading microorganisms and a key player in garbage disposal systems throughout the body. Through a tightly coordinated cascade of events, complement mediates pathogen recognition and destruction either via opsonisation followed by phagocytosis or by the formation of a lytic pore, the membrane attack complex (MAC). These processes are accompanied by the production of anaphylatoxins, C3a and C5a, which drive inflammation and facilitate immune cell recruitment.

Complement can be activated through one of three pathways; classical (CP), lectin (LP), or alternative (AP) (Figure 1). CP activation is initiated by binding of the C1q/r/s complex to a surface either via surface-bound antibodies (IgG and IgM) or a variety of self-molecules such as Aβ and apoptotic markers including phosphatidylserine and extracellular DNA [5,6]; binding triggers activation of C1s, a serine protease that cleaves C4 and C2 to produce the membrane-bound C3-convertase, C4b2a. The LP is activated by mannose-binding lectin (MBL) or ficolin, these bind carbohydrate epitopes on surfaces. MBL-associated serine proteases (MASPs) cleave C4 and C2 to generate C4b2a, as in the CP. The AP is better considered as an amplification loop whereby either spontaneously hydrolysed C3 ($C3(H_2O)$) or C3b generated in the CP/LP, bind factor B (FB), catalysing FB cleavage by factor D (FD) to form the AP C3-convertase ($C3(H_2O)Bb$ or C3bBb), which cleaves more C3 to generate membrane-bound opsonin C3b. The AP loop is therefore self-perpetuating and rapidly activating, critical for successful pathogen clearance, but dysregulation can be extremely costly. The three pathways converge at the point of C3 cleavage; each C3-convertase cleaves multiple C3 molecules into C3a and C3b leading to widespread complement deposition. C3b binding adjacent to C3-convertases creates the C5-convertases C4b2a3b and C3bBb3b, which cleave C5 into C5a and C5b. C3a and C5a are potent proinflammatory anaphylatoxins that recruit and activate immune cells expressing C3a and C5a receptors. C5b sequentially recruits C6, C7, C8, and C9 to form the membrane attack complex (MAC), which through a series of conformational changes, punches through the cell membrane resulting in cell lysis or cell activation.

To avoid damage to self, complement is tightly controlled at every level of the pathway by an array of regulators in fluids and on cell surfaces (Figure 1). Nevertheless, overactivation or failure of complement regulators to keep the pathway in check can trigger a vicious cycle of inflammation and tissue damage.

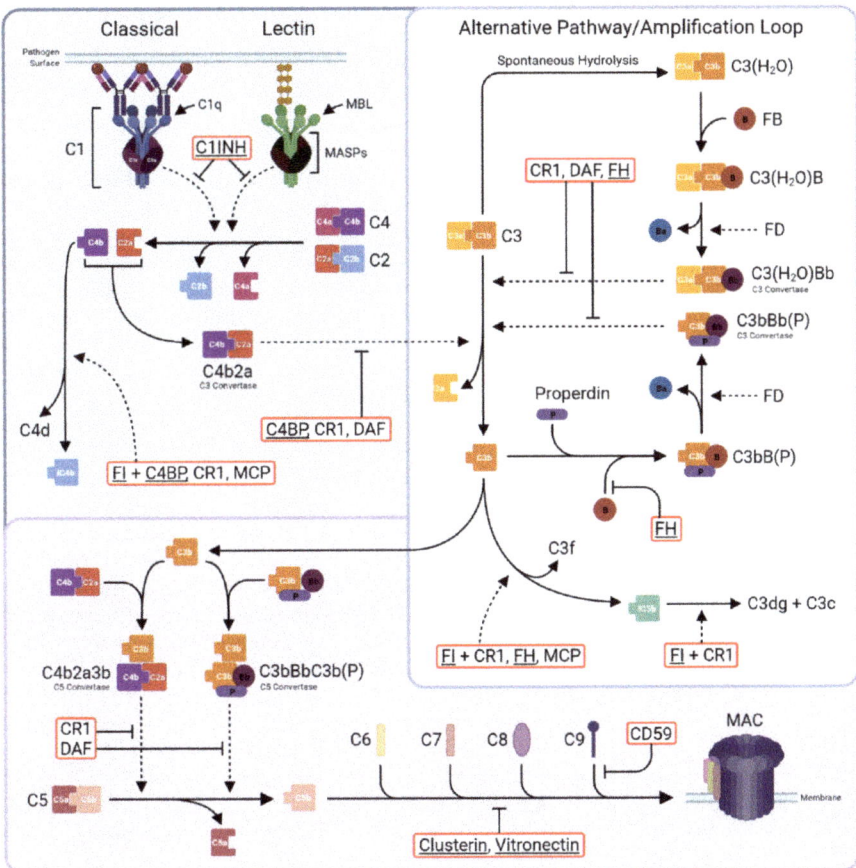

Figure 1. The complement system. Three activation pathways converge on the central C3 molecule. The classical pathway is triggered by binding of antibody-antigen complexes to C1 via C1q subunits. C1r proteolytically activates C1s, which in turn cleaves C4 and C2 to form the classical C3 convertase C4b2a. The lectin pathway begins with recognition of pathogen surface carbohydrates by mannose-binding lectin (MBL) followed by activation of MBL-associated serine proteases (MASPs), which also cleave C4 and C2 to generate C4b2a. The alternative pathway is an amplification loop initiated by C3b generated in the above activation pathways or by spontaneous hydrolysis of C3 to $C3(H_2O)$. Factor B (FB) then binds $C3b/C3(H_2O)$, enabling its cleavage by Factor D (FD) to form the alternative pathway C3 convertase $C3bBb/C3(H_2O)Bb$; binding of properdin (P) stabilises the convertase. Both C3 convertases cleave C3 into C3a and C3b. The classical and lectin pathways are negatively regulated by C1-inhibitor (C1INH), which inhibits both C1s and MASPs, while the C3 convertases are regulated by C4b-binding protein (C4BP; specific for C4b2a), decay-accelerating factor (DAF; specific for C3bBb), complement receptor 1 (CR1), and Factor H (FH), either directly through increasing decay or indirectly by catalysing cleavage of C4b by Factor I (FI). At the next stage of the pathway, C3b is incorporated into the C3 convertases to form the C5 convertases C4b2a3b and C3bBbC3b(P). These are regulated in the same manner as the C3 convertases and cleave C5 into C5a and C5b to trigger the terminal pathway. C5b is sequentially bound by C6, C7, C8, and up to 18 C9 molecules to form the membrane attack complex (MAC); MAC assembly is inhibited by clusterin and vitronectin in the fluid phase and CD59 on cells. Complement regulators are in red boxes, fluid-phase regulators are underlined. Solid, dotted, and blunt arrows indicate pathway progression, proteolytic cleavage, and direct inhibition, respectively.

2. Genetics Implicate Inflammation, Immunity, and Complement in the Pathogenesis of Late-Onset AD

Late-onset AD (LOAD), responsible for ~95% of AD cases, is a multifactorial disease with a heritability of over 58% [7]. Since 2009, large genome-wide association studies (GWAS) have identified over 75 independent genetic risk factors for LOAD [3,8–10]. In silico pathway analyses have implicated amyloid and tau processing, lipid, and innate immunity pathways [4]. Approximately 20% of LOAD risk loci encode proteins implicated in immunity; many of these have roles in macrophage and microglial activation, an observation supported by recent single-cell expression enrichment analyses [8]. Among the GWAS statistically significant (GWS) hits are two genes encoding proteins of the complement pathway; *CR1* encoding the membrane protein complement receptor 1 (CR1) and *CLU* encoding the plasma regulator clusterin. Additionally, *C1S* encoding the enzyme C1s reaches near GWS in the most recent GWAS [8]. *CR1* and *CLU* are among the most significant GWAS hits, ranking high in the top 10. These strong associations provide the impetus for this review of complement genetics in LOAD.

3. Complement Genetic Variation Impacts Risk of Inflammatory Disease

Genetic variations within complement genes are extremely widespread in the general population; over the last 20 years, many common polymorphisms and rare mutations in complement genes have been linked with diverse inflammatory and infectious diseases, demonstrating the pivotal role of the complement pathway in determining disease risk (Table 1). Occasionally, these genetic variants are the primary cause of a disease through either causing deficiency or significant gain or loss of function changes in complement components or regulators; more commonly, functional changes associated with variants are subtle and exacerbate existing pathology by contributing to a vicious cycle of inflammation and tissue damage.

Considering the common polymorphisms, individual variants usually have only minor effects on protein function and complement homeostasis, but the additive effects of combinations of variants in multiple complement genes can have major effects, tipping the balance in favour of complement dysregulation and impacting disease predisposition. The combination of common genetic variants in complement genes that defines the complement genetic make-up of an individual is referred to as the "complotype" [11].

The complotype has been best studied in the context of age-related macular degeneration (AMD), progressive retinal disease, and the leading cause of blindness in the developed world. Common variants in genes encoding the AP components C3 and FB and the AP regulator FH are individually associated with higher C3 convertase activity and increased AMD risk; a combination of risk variants in these three genes (*C3* (rs2230199), *CFB* (rs641153), and *CFH* (rs800292)) increased complement activity in plasma six-fold [12]. This complotype, and another *CFB* variant (rs4151667), were later associated with AMD disease status and increased complement activation markers (C3d/C3 ratio) in AMD plasma [13]. These variants were also associated with an increased risk of dense deposit disease (DDD), a renal disease characterised by systemic AP activation and complement deposition in the kidneys. In contrast, the AP gene variants conferring risk for AMD and DDD were not risk variants for another renal disease associated with complement dysregulation, atypical haemolytic uremic syndrome (aHUS), a disease characterised by thrombocytopenia, microangiopathic haemolytic anaemia, and acute renal failure with complement deposition in the kidney [14]. This lack of concordance of risk suggests that the roles of complement are quite different in these superficially similar diseases; in support of this, a common genetic variation that causes deletion of the genes encoding FH-related proteins 1 and 3 (*CFHR1/CFHR3*) is protective for AMD but increases the risk of aHUS [14]. These findings demonstrate that the same complement gene variant, or set of variants, can be involved in several diseases and that specific variants may have inverse effects on risk in some apparently similar diseases. Better knowledge of the effects of these variants on complement regulation in plasma and in tissues will inform understanding of mechanisms of disease.

Table 1. Complement gene variants and associated diseases.

Gene	Variant	Disease
C1q	Deficiency Polymorphism	Increased risk of lupus and glomerulonephritis Arthritis, cancer, diabetes, schizophrenia
C1r/C1s	Deficiency GOF SNP	Autoimmunity, infections, glomerulonephritis, Type I periodontal Ehlers-Danlos Increased risk of AD
C1INH	Deficiency	Hereditary angioedema (types I and II)
C2	Deficiency SNPs	Lupus, bacterial infections Protective for AMD and PCVP
C3	GOF Nonsynonymous Coding variant	aHUS, C3G, and AMD
C4	Deficiency CNV	Lupus Schizophrenia
C5	Nonsense; hom or Compound het	C5 deficiency; neisserial infections
C6	Single bp deletion	C6 deficiency; neisserial infections
C7	Nonsense: hom or compound het	C7 deficiency; neisserial infections
C8α	Nonsense: hom or compound het	C8 deficiency, type I; neisserial infections; no C8α protein; free C8β
C8β	Premature stop codon	C8 deficiency, type II; neisserial infections; no C8β protein; free C8α
C9	Nonsense: hom or compound het SNPs	C9 deficiency; neisserial infections AMD; AD
MASP-1, collectins	Hom/het deficiency	Various developmental; Malpuech, Carnevale, Michels, and Mingarelli syndrome
Ficolins	SNPs	Rheumatoid arthritis, leprosy, systemic inflammation, bacterial infections
CFH	Hom deficiency SNPs and truncations	DDD; MPGN C3G; acquired partial lipodystrophy; aHUS AMD; AD; Some protective against meningococcal disease, AMD, IgAN, or C3G
CFI	Nonsense: hom, het or compound het	AMD; C3G; aHUS; recurrent infections
MCP	Hom/Het deletion/truncation Missense SNP	Systemic sclerosis, miscarriage, HELLP syndrome, and C3G Severe aHUS; linked to CVID
CFB	Nonsense: hom or compound het Het GOF SNP Other SNPs	Factor B deficiency; recurrent bacterial infections aHUS Protection against AMD
Properdin	Nonsense/truncating mutations	Properdin deficiency (X-linked); neisserial infections
DAF	Nonsense: hom or compound het	CHAPLE Syndrome; linked to Inab Cromer blood group
CD59	Nonsense: hom or compound het	CD59 deficiency; PNH-like disease; Peripheral neuropathy; strokes
CFHR1/3	Combined gene deletion	Risk for aHUS; protection from AMD
CFHR5	Gene duplication SNPs	aHUS C3G; poststreptococcal glomerulonephritis
Clu	SNPs	AD
CR1	SNPs	AD

AD—Alzheimer's disease, aHUS—atypical haemolytic uremic syndrome, AMD—age-related macular degeneration, bp—basepair, C3G—complement 3 glomerulopathy, CHAPLE—complement hyperactivity, angiopathic thrombosis, and protein-losing enteropathy, CNV—copy number variant, CVID—common variable immunodeficiency, DDD—dense deposit disease, GOF—gain of function, het—heterozygous, hom—homozygous, MPGN—membranoproliferative glomerulonephritis, LOF—loss of function, PCVP—polypoidal choroidal vasculopathy, PNH—paroxysmal nocturnal hemoglobinuria, SNP—single nucleotide polymorphism.

4. Complement in LOAD

In post-mortem analyses of LOAD brain, complement components and activation products, notably C1q, C4b, C3b/iC3b, and MAC, co-localise with amyloid plaques and neurofibrillary tangles [5,15–17]. By default, these studies only address late/end-stage disease and provide no clues as to how complement activation impacts the disease. Given the role of complement in "taking out the trash", one likely role of complement in LOAD is in facilitating the removal of accumulated amyloid plaques and tangles, dead and dying cells. Indeed, Aβ peptides, the precursors of amyloid, when exposed to serum, activate both the CP and AP and are opsonised by C3b/iC3b fragments [18]; this would enable recognition and phagocytosis by cells expressing complement receptor CR3, including CNS resident microglia (Figure 2). Outside of the brain, C3b-opsonised Aβ aggregates can bind CR1 on erythrocytes, a pathway for clearance in the liver [19]. These findings suggest that complement activation may have a protective role in early disease, provoking local phagocytosis of amyloid by resident cells and peripheral clearance; however, complement is a double-edged sword, protective when properly regulated but with the potential to cause damage when dysregulated. Dysregulated complement can then drive inflammation and directly activate or damage self-cells. Importantly, complement activation has been implicated in synapse pruning and loss, both physiological during brain development and pathological in neurodegeneration [20–23]. C1 tags synapses destined for removal and trigger CP activation leading to deposition of opsonic C3 fragments, signalling microglial phagocytosis. The demonstration that mice deficient in C1q or C3 show reduced synapse loss emphasises the importance of this process [23].

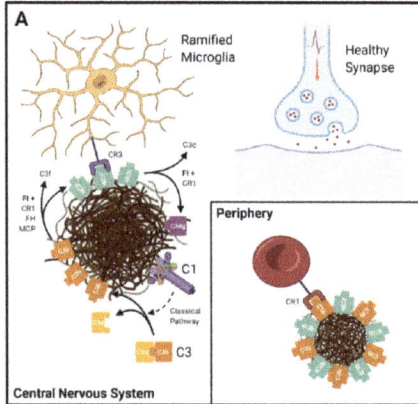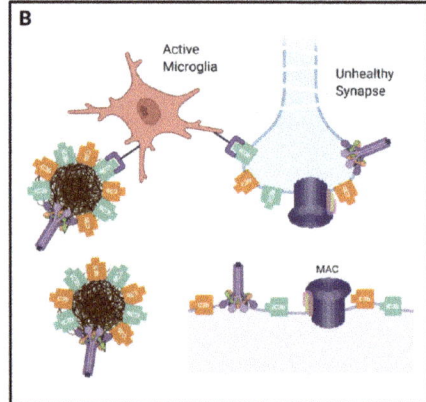

Figure 2. The Janus-faced nature of complement in AD: (**A**) In the central nervous system complement components and activation products (C1q/r/s and C3b) are deposited on amyloid plaques. C3b is converted to iC3b by Factor I (FI) with cofactor activity from CR1, Factor H, or MCP. iC3b binds to phagocytic receptor CR3 (an integrin dimer comprising CD11b and CD18 chains) on the surface of microglia, enabling plaque clearance. iC3b is further broken down by FI and CR1 into inactive C3dg. In the periphery, CR1 binds to C3b-opsonised amyloid aggregates and transports them to the liver for destruction in a process called "immune complex clearance". (**B**) Complement dysregulation tips the balance towards destruction. In the absence of proper CR1 function, complement components accumulate, resulting in cell activation or damage. Complement is also involved in pathological synapse loss in AD. C1 binds to a poorly defined receptor on synapses and triggers classical pathway activation, resulting in C3b opsonisation and subsequent phagocytosis by activated microglia.

Whether complement activation is beneficial or detrimental for LOAD progression depends on regulation. Inappropriate activation or dysregulation of complement will drive pathological inflammation and has been implicated in inflammatory brain diseases such as neuromyelitis optica and multiple sclerosis [24,25]. The strongest evidence implicating complement in LOAD aetiology comes from genetic studies; genome-wide association studies (GWAS) implicated *CR1* and *CLU*, respectively encoding the complement receptor

CR1 and the fluid-phase regulator clusterin [3,9,26]; the most recent LOAD GWAS reported a novel suggestive association of *C1S* the gene encoding the critical CP enzyme C1s, with risk [8]. Below we will briefly describe each of these complement hits, address the nature of their LOAD associations and explore mechanisms.

5. CR1

5.1. Function

CR1 is a receptor for the complement activation products C3b and C4b and a number of other ligands, detailed below. Once bound to CR1, C3b and C4b can be cleaved by the plasma protease FI, with CR1 itself providing the essential cofactor activity. The cleavage products (iC3b and C4c, respectively) have a minimal affinity for CR1; this binding-cleavage-release cycle is critical for the role of CR1 in immune complex (IC) handling [27]. C3b/C4b-coated ICs bind CR1 on erythrocytes in the circulation and are ferried to the liver and spleen for transfer to tissue macrophages expressing CR3 (the receptor for iC3b, now abundant on the IC) for phagocytic elimination. CR1 also has decay-accelerating activity for the C3 and C5 convertases; it binds C4b displacing C2a and binds C3b displacing Bb; this capacity to decay CP and AP convertases confers powerful complement regulating activity, although this is likely of minor physiological importance.

5.2. Expression

CR1 is expressed on erythrocytes where it performs the critical IC transport role described above; indeed, reduced CR1 levels on erythrocytes is strongly associated with the immune complex disease systemic lupus erythematosus (SLE), although whether this is cause or effect remains a subject of debate [28]. CR1 is also expressed on leukocytes in blood (neutrophils, monocytes, B cells), on macrophages and dendritic cells in tissues, and on podocytes in the kidney. In the brain, CR1 expression has been demonstrated in neurons and astrocytes in post-mortem LOAD and multiple sclerosis brain tissue [29–31]. CR1 expression has also been reported in cultured primary human astrocytes and microglia, and on human stem cell-derived microglia transplanted into mouse brain [31–33]; however, there is a continuing debate with some suggesting that CR1 is not expressed in the brain and that the impact of CR1 on AD is explained by its peripheral roles in IC handling [34]. A clear understanding of whether, where, and when CR1 is expressed in the brain is essential for our understanding of how *CR1* single nucleotide polymorphisms (SNPs) might confer increased LOAD risk.

5.3. Structure and Genetic Variants

The *CR1* gene is located on chromosome 1q32 within the regulators of complement activation (RCA) gene cluster; like other members of this cluster, it is a highly repetitive gene made up of repeating units with internal duplications that cause copy number variation (CNV). CNV in *CR1* generates four co-dominant alleles that encode CR1 proteins differing in the number of long homologous repeats (LHRs) (Figure 3). CR1*1 (also called CR1-A or CR1-F), a 190 kDa protein, is the most common variant with an allele frequency of 0.87; it comprises four LHRs, each made up of seven short consensus repeats (SCRs; 60–70 amino acid, internally disulphide-bonded structural units), an additional two membrane-proximal SCRs, transmembrane and cytoplasmic regions. CR1*2 (also called CR1-B or CR1-S) has an extra LHR, a duplication of SCR 3–9, yielding a 220 kDa protein; it has an allele frequency of 0.11. The remaining alleles, CR1*3 (also called CR1-C or CR1-F'; 160 kDa) and CR1*4 (also called CR1-D; 250 kDa), are very rare [35,36]. CR1*2 increases risk of LOAD by ~30% [3,9,30,37]. The addition of an extra LHR in CR1*2 increases the number of C3b/C4b binding sites, a theoretical gain of function (Figure 3) [38,39]. The increased risk associated with a gain-of-function variant in a molecule essential for IC clearance is counter-intuitive; one plausible explanation is that expression of the CR1*2 haplotype is reduced; indeed, reduced CR1 expression on erythrocytes in CR1*2 carriers has been reported [40,41]. It has been suggested that the expression of CR1*2 is reduced compared to CR1*1 because it is less efficiently trafficked to the membrane, remaining trapped in cytoplasmic vesicles [30]. Whether the CR1*2 allele is associated with a reduced expression on CNS resident immune cells remains to be demonstrated.

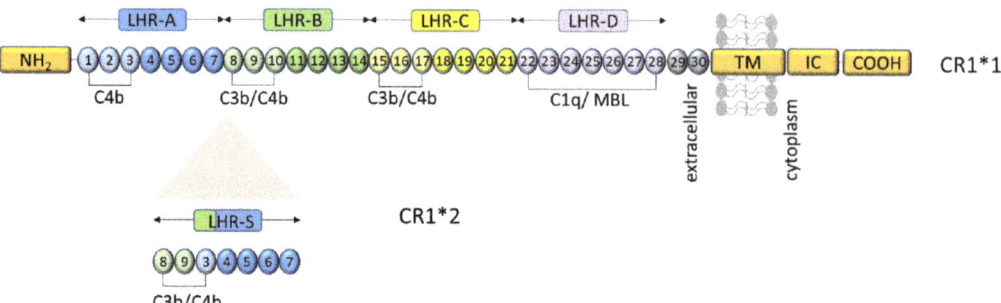

Figure 3. Representation of CR1 structure and ligand binding sites. CR1*1 comprises, from the amino terminus (NH$_2$), four long homologous repeats (LHRs A-D), each composed of seven short consensus repeats (SCRs) of 60–70 amino acids each, two additional SCRs, a transmembrane segment (TM), and an intracytoplasmic carboxy-terminal domain (IC-COOH). Each circular block represents an SCR (numbered 1–30). There are three C4b binding sites (SCR 1–3, 8–10, and 15–17) and two C3b binding sites (SCR 8–10 and 15–17). SCRs 22–28 bind C1q, MBL, and ficolins. CR1*2 has an additional LHR domain (LHR-S) inserted between LHRs A and B and consequently an extra C3b/C4b binding site. Schematic based on similar figures in the work of [37,39,42].

The most recent meta-analysis of LOAD GWAS identified rs679515 as the most significant *CR1* risk SNP [8]. Prior to this, rs4844610 and rs6656401 were reported [3,9]. All three SNPs are intronic, and all are in linkage disequilibrium. This SNP association marks the CNV described above, providing a means of identifying risk CNV carriers and clues to the mechanism [37,38]. A single rare coding variant, rs4844609, has been identified that is associated with episodic memory [43]. This SNP causes a Ser1610Thr substitution at a membrane-proximal site in LHR-D of CR1 previously implicated as a C1q binding site [44]. One study reported that the risk variant at this SNP increased the binding affinity of CR1 for C1q [31]; however, this was not replicated using recombinant CR1 LHR-D containing this Ser/Thr substitution [44]. Others suggested that the Ser1610Thr change altered susceptibility to enzymatic cleavage of CR1 and generation of soluble CR1 (sCR1), a locally active, fluid-phase complement inhibitor that might impact dysregulation of complement in the surrounding milieu. Indeed, increased plasma levels of sCR1 have been associated with both rs4844609 and rs6656401 [31,40]. It was suggested that rs4844609 accounts for the known LOAD risk effect of rs6656401 [43]; however, this has been refuted by others [45]. To date, the LOAD-associated SNPs in *CR1* were identified from GWAS in Caucasian populations [3,9,46–48]. The few analyses of non-European populations have reported conflicting results, some reports showing association of these same variants in *CR1* with LOAD in, for example, Han Chinese populations [49,50], whereas others failed to replicate the findings from Caucasian populations [51].

6. Clusterin

6.1. Function

Clusterin is a multifaceted protein; its many and diverse functions were discovered independently of each other; hence, clusterin has many names in the literature [52]. Clusterin is a lipoprotein that, in addition to roles in lipid transport, is an extracellular chaperone with roles in BAX-mediated apoptosis, PI3K pro-survival, and oxidative stress pathways [53–56]. Clusterin also contributes to the regulation of the complement system; it is a fluid-phase inhibitor of the terminal complement pathway, binding MAC precursors in the fluid phase to prevent membrane binding and pore formation [57,58].

6.2. Expression

Clusterin is ubiquitously expressed in tissues. Alternative splicing generates three forms of clusterin that are, respectively, nuclear, cytoplasmic, and secreted. The first two are regulators of apoptosis and intracellular chaperones and are not discussed further here. Secreted clusterin is present in plasma at a concentration of ~100 mg/L; a proportion of this will be contained within lipoprotein particles. Clusterin is also present in cerebrospinal fluid (CSF) and other biological fluids, notably at high levels in seminal plasma. Clusterin is abundantly expressed in the CNS, predominantly by astrocytes with region-specific expression in a subset of neurons [59,60]. In the healthy brain, astrocytes are responsible for the production and secretion of clusterin into the extracellular space. Overexpression of both neuronal and astrocytic clusterin has been reported in cases of inflammatory insult and neurodegenerative disease, including traumatic brain injury and spinal cord injury [61–64].

A role for clusterin in LOAD was first reported over 30 years ago. Clusterin mRNA is upregulated in AD tissue [65], and clusterin protein is abundant in the AD brain, where it is found in a subset of plaques and co-localises with MAC-labelled dystrophic neurites, neuropil threads, amyloid deposits, and intracellular neurofibrillary tangles [66,67]. Clusterin expression positively correlated with *ApoE4* allele number [68]. Levels of clusterin are elevated in the CSF and plasma of LOAD patients [69,70]; indeed, plasma clusterin has been suggested as a biomarker for AD, correlating with disease severity and progression from mild cognitive impairment (MCI) to AD in some studies [71–73]. Precisely how clusterin impacts the pathogenesis of LOAD remains unclear. In an in vitro acellular system, clusterin prevented Aβ aggregation [74]. Clusterin and the Clu-receptor glycoprotein 330/megalin have been reported to complex with soluble Aβ (sAβ) in the brain in order to facilitate the transport of sAβ across the blood-brain-barrier (BBB) [75]. Others have shown that clusterin binds and sequesters Aβ_{1-40} aggregates in vitro [76]. In mouse models, $Clu^{-/-}ApoE^{-/-}$ double knockout mice showed markedly increased Aβ production and amyloid deposition compared with either single knockout, suggesting cooperative effects of these lipoproteins [77,78]. Recent studies have also suggested a role for clusterin at the synapse with increased clusterin protein reported in synaptoneurosomes from AD patients and in *ApoE4* carriers [79].

6.3. Structure and Genetic Variants

Clusterin is a heavily glycosylated heterodimeric protein comprising α and β chains each of ~40 kDa molecular weight, generated from an 80 kDa precursor protein and linked by five disulphide bonds. The structure is poorly defined, in part because of its tendency to aggregate; however, both chains contain stretches of amphipathic helix interspersed with disordered regions. The resultant molecule is highly flexible, likely explaining its broad range of binding partners. The gene encoding clusterin (*CLU*) is found on chromosome 8p21-12 and comprises nine exons. The primary transcript (NM_001831.3) encodes an immature pre-pro-protein containing a 22 amino acid signal sequence for translocation to the endoplasmic reticulum (ER). At the ER, immature clusterin is processed and cleaved to yield the highly glycosylated, mature heterodimeric protein.

Rare nonsynonymous mutations in *CLU* have been reported in a subset of AD patients and shown to result in intracellular accumulation of CLU in the ER and loss of secreted clusterin at the Golgi apparatus [80]. Of more relevance, there is an abundance of genetic evidence associating variants within the *CLU* gene with increased LOAD risk; indeed, *CLU* is the third strongest genetic risk factor for LOAD to date. Independent studies have identified multiple SNPs in *CLU*, associated with increased LOAD risk (rs11136000, rs2279590, rs9331888, rs9331896 and rs11787077) [8,9,26]. To date, there is no clear mechanism to explain how these clusterin variants confer increased LOAD risk, a task that is greatly complicated by the promiscuity of the protein. Whether and how SNPs in *CLU* affect clusterin synthesis systemically and locally in the CNS remains to be determined.

All SNPs studied to date have been suggested to affect plasma clusterin levels [81–83]. The rs11136000 SNP is located in intron 3 of *CLU*; 88% of Caucasians carry the C allele, this increases LOAD risk 1.6-fold [9]. The C allele is also associated with the risk of mild cognitive impairment (MCI) and progression from MCI to AD [83,84]. The minor T allele shows a mild protective effect [85]; this SNP has also recently been shown to be associated with cognitive decline in Parkinson's disease patients [86]. The rs9331888 risk SNP has been associated with low levels of plasma clusterin and linked to alternative splicing of the *CLU* gene [82,87,88]. It should be stressed that *CLU* may impact LOAD risk independently of complement regulation via its roles in lipid handling and Aβ clearance; this has been expertly reviewed elsewhere [89].

7. C1S

7.1. Function

C1s is a single-chain glycoprotein, a highly specific serine protease, and a core component of the C1 complex, the initiator of the classical complement pathway. C1q is the recognition unit of the complex, binding antibody or other ligands; conformational changes in C1q then activate the associated pro-enzyme C1r, which in turn proteolytically activates pro-C1s. Activated C1s can then cleave C4 and C2 to form the C3 convertase C4b2a. Deficiency of C1s (or any of the components of the C1 complex) is strongly associated with a lupus-like immune complex disease reflecting loss of capacity to activate complement on immune complexes.

7.2. Expression

The *C1S* gene is located on chromosome 12, where the *C1R* and *C1S* genes lie end to end separated by 9.5 kb; they are derived from a common ancestral gene through gene reduplication [90]. C1s are predominantly made in hepatocytes but are also produced by activated macrophages and monocytes. Brain expression is low and predominantly by microglia [91]. The plasma concentration of C1s is ~30 mg/L, the large bulk of this incorporated in the C1 complex. C1s are also present in CSF, although absolute levels were not obtained [92].

7.3. Genetic Variants

Complete C1s deficiency is associated with the immune complex disease as noted below; partial deficiencies have been associated with Ehlers-Danlos syndrome though the underlying mechanisms are unclear. Until very recently, no other disease-associated variants in *C1S* were reported. The most recent GWAS identified a novel SNP 5Kb upstream of *C1S*, which showed suggestive association with increased LOAD risk (SNP rs3919533) [8]. The mechanism of action of this SNP remains to be determined through functional experiments; however, given the location of the SNP, it is likely to impact the expression of the protein; indeed, C1s levels have previously been shown to be reduced in the CSF of AD patients, though there is no evidence that this observation is related to the *C1S* risk SNP [92].

8. Complement in LOAD: Smoking Gun or Red Herring?

In many chronic inflammatory and degenerative diseases, a role for complement has been clearly demonstrated, often with evidence pinpointing the relevant parts of the complement pathway involved in disease aetiology, for example, alternative pathway activation in AMD, and sometimes with proven efficacy of anti-complement drugs. Until very recently, the situation for LOAD was very different; complement proteins and activation products had been demonstrated in LOAD brains and biological fluids, but this "guilt-by-association" was not supported by solid evidence. Two things have changed the situation; first, the demonstration that complement activation at the synapse is a critical player in synapse loss in the disease; second, the genetic evidence implicating complement summarised above. The genetics tell us that *CR1*, *CLU*, and likely *C1S* are strongly implicated in the disease process—although whether the clusterin association involves its complement roles is very unclear. While this provides strong evidence that complement dysregulation is involved in LOAD, it does not point towards a specific pathway or mechanism.

Understanding how complement variants confer LOAD risk is further complicated by several factors. Firstly, the majority of the LOAD-associated complement variants identified to date are non-coding and likely confer risk by affecting cell and region-specific expression levels. Unlike in other more accessible organs, the location and nature of the brain make it impossible to assess longitudinal expression levels in the brain parenchyma, and reliance on post-mortem evidence likely masks important early and progressive changes. Second, LOAD-associated complement genes predispose individuals to LOAD, but other risk factors (non-complement and non-genetic) are required to cause disease; hence, functional studies of risk variants must be conducted in specific contexts to reveal relevant mechanistic pathways of action. In vivo and in vitro studies, each with different limitations must be used in conjunction to understand the role of complement at different time points in disease. Third, there are many regions of the human genome, including some important complement loci, which cannot be assembled or aligned using standard short-read sequencing technologies, preventing the identification of disease-causing mutations or variations [93,94]. These regions are referred to as "dark" or "camouflaged"; "dark" regions are difficult to sequence due to, for example, high GC content, while "camouflaged" regions of the genome are highly repetitive, making alignment of short reads difficult. The complexity of complement genes is a consequence of gene reduplication events, so many loci, notably the RCA cluster, are highly repetitive in nature and therefore likely well camouflaged. For example, regarding *CR1* in the RCA cluster, 26% of the protein-coding region is hidden due to its highly repetitive nature so that significant variation may be missed by standard sequencing in GWAS [94]. Indeed, this study, systematically targeting "dark" genes relevant to LOAD risk, identified a novel 10-nucleotide frameshift mutation in CR1 present in five cases but no controls.

Our recent study using available GWAS data identified no remaining complement gene LOAD association when *CLU* and *CR1* were removed from a complement geneset [95]; however, such analyses are limited by the data. Indeed, the recent GWAS identification of a suggestive association of *C1S* with LOAD [8] highlights that larger data sets and newer sequencing technologies may identify other complement genes that impact LOAD risk.

Often, by the time people with LOAD reach the clinic, they already have significant irreversible pathology. Understanding of complement risk genes and the resultant complotypes involved in LOAD might facilitate predictive genetic screening tests; if the complotypes can be linked with complement levels in plasma, as seen in AMD, this might enable the development of simple diagnostic tools and guide the future use of anti-complement drugs in LOAD. There are a number of anti-complement therapeutics currently in development for CNS disorders [96]; genetic and biomarker assays could be used to stratify patients for anti-complement therapeutic interventions.

Author Contributions: Writing—original draft preparation, M.T., S.M.C., and B.P.M.; writing—review and editing, M.T., S.M.C., N.D., R.A.J.B., R.S., and B.P.M. All authors have read and agreed to the published version of the manuscript.

Funding: This work is supported by the U.K. Dementia Research Institute, which receives its funding from U.K. DRI Ltd., funded by the U.K. Medical Research Council, Alzheimer's Society, and Alzheimer's Research U.K.

Institutional Review Board Statement: Not applicable.

Informed Consent Statement: Not applicable.

Acknowledgments: Figures 1 and 2 created with BioRender.com.

Conflicts of Interest: The authors declare no conflict of interest.

References

1. World Health Organization. *Towards a Dementia Plan: A WHO Guide*; World Health Organization: Geneva, Switzerland, 2018.
2. Kinney, J.W.; Bemiller, S.M.; Murtishaw, A.S.; Leisgang, A.M.; Salazar, A.M.; Lamb, B.T. Inflammation as a central mechanism in Alzheimer's disease. *Alzheimers Dement. N. Y.* **2018**, *4*, 575–590. [CrossRef]
3. Kunkle, B.W.; Grenier-Boley, B.; Sims, R.; Bis, J.C.; Damotte, V.; Naj, A.C.; Boland, A.; Vronskaya, M.; van der Lee, S.J.; Amlie-Wolf, A.; et al. Genetic meta-analysis of diagnosed Alzheimer's disease identifies new risk loci and implicates Abeta, tau, immunity and lipid processing. *Nat. Genet.* **2019**, *51*, 414–430. [CrossRef]
4. Jones, L.; Holmans, P.A.; Hamshere, M.L.; Harold, D.; Moskvina, V.; Ivanov, D.; Pocklington, A.; Abraham, R.; Hollingworth, P.; Sims, R.; et al. Genetic evidence implicates the immune system and cholesterol metabolism in the aetiology of Alzheimer's disease. *PLoS ONE* **2010**, *5*, e13950. [CrossRef] [PubMed]
5. Rogers, J.; Cooper, N.R.; Webster, S.; Schultz, J.; McGeer, P.L.; Styren, S.D.; Civin, W.H.; Brachova, L.; Bradt, B.; Ward, P.; et al. Complement activation by β-amyloid in Alzheimer disease. *Proc. Natl. Acad. Sci. USA* **1992**, *89*, 10016–10020. [CrossRef]
6. Jiang, H.; Cooper, B.; Robey, F.A.; Gewurz, H. DNA binds and activates complement via residues 14–26 of the human C1q A chain. *J. Biol. Chem.* **1992**, *267*, 25597–25601. [CrossRef]
7. Gatz, M.; Reynolds, C.A.; Fratiglioni, L.; Johansson, B.; Mortimer, J.A.; Berg, S.; Fiske, A.; Pedersen, N.L. Role of genes and environments for explaining Alzheimer disease. *Arch. Gen. Psychiatry* **2006**, *63*, 168–174. [CrossRef] [PubMed]
8. Bellenguez, C.; Küçükali, F.; Jansen, I.; Andrade, V.; Moreno-Grau, S.; Amin, N.; Naj, A.C.; Grenier-Boley, B.; Campos-Martin, R.; Holmans, P.A.; et al. New insights on the genetic etiology of Alzheimer's and related dementia. *medRxiv* **2020**. [CrossRef]
9. Lambert, J.C.; Heath, S.; Even, G.; Campion, D.; Sleegers, K.; Hiltunen, M.; Combarros, O.; Zelenika, D.; Bullido, M.J.; Tavernier, B.; et al. Genome-wide association study identifies variants at CLU and CR1 associated with Alzheimer's disease. *Nat. Genet.* **2009**, *41*, 1094–1099. [CrossRef]
10. Sims, R.; van der Lee, S.J.; Naj, A.C.; Bellenguez, C.; Badarinarayan, N.; Jakobsdottir, J.; Kunkle, B.W.; Boland, A.; Raybould, R.; Bis, J.C.; et al. Rare coding variants in PLCG2, ABI3, and TREM2 implicate microglial-mediated innate immunity in Alzheimer's disease. *Nat. Genet.* **2017**, *49*, 1373–1384. [CrossRef]
11. Harris, C.L.; Heurich, M.; Rodriguez de Cordoba, S.; Morgan, B.P. The complotype: Dictating risk for inflammation and infection. *Trends Immunol.* **2012**, *33*, 513–521. [CrossRef]
12. Heurich, M.; Martinez-Barricarte, R.; Francis, N.J.; Roberts, D.L.; Rodriguez de Cordoba, S.; Morgan, B.P.; Harris, C.L. Common polymorphisms in C3, factor B, and factor H collaborate to determine systemic complement activity and disease risk. *Proc. Natl. Acad. Sci. USA* **2011**, *108*, 8761–8766. [CrossRef]
13. Paun, C.C.; Lechanteur, Y.T.E.; Groenewoud, J.M.M.; Altay, L.; Schick, T.; Daha, M.R.; Fauser, S.; Hoyng, C.B.; den Hollander, A.I.; de Jong, E.K. A Novel Complotype Combination Associates with Age-Related Macular Degeneration and High Complement Activation Levels in vivo. *Sci. Rep.* **2016**, *6*, 26568. [CrossRef]
14. De Cordoba, S.R.; Tortajada, A.; Harris, C.L.; Morgan, B.P. Complement dysregulation and disease: From genes and proteins to diagnostics and drugs. *Immunobiology* **2012**, *217*, 1034–1046. [CrossRef] [PubMed]
15. Ishii, T.; Haga, S. Immuno-electron-microscopic localization of complements in amyloid fibrils of senile plaques. *Acta Neuropathol.* **1984**, *63*, 296–300. [CrossRef]
16. Veerhuis, R.; van der Valk, P.; Janssen, I.; Zhan, S.S.; Van Nostrand, W.E.; Eikelenboom, P. Complement activation in amyloid plaques in Alzheimer's disease brains does not proceed further than C3. *Virchows Arch.* **1995**, *426*, 603–610. [CrossRef]
17. Webster, S.; Lue, L.F.; Brachova, L.; Tenner, A.J.; McGeer, P.L.; Terai, K.; Walker, D.G.; Bradt, B.; Cooper, N.R.; Rogers, J. Molecular and cellular characterization of the membrane attack complex, C5b-9, in Alzheimer's disease. *Neurobiol. Aging* **1997**, *18*, 415–421. [CrossRef]
18. Bradt, B.M.; Kolb, W.P.; Cooper, N.R. Complement-dependent proinflammatory properties of the Alzheimer's disease β-peptide. *J. Exp. Med.* **1998**, *188*, 431–438. [CrossRef]
19. Rogers, J.; Li, R.; Mastroeni, D.; Grover, A.; Leonard, B.; Ahern, G.; Cao, P.; Kolody, H.; Vedders, L.; Kolb, W.P.; et al. Peripheral clearance of amyloid β peptide by complement C3-dependent adherence to erythrocytes. *Neurobiol. Aging* **2006**, *27*, 1733–1739. [CrossRef] [PubMed]
20. Stevens, B.; Allen, N.J.; Vazquez, L.E.; Howell, G.R.; Christopherson, K.S.; Nouri, N.; Micheva, K.D.; Mehalow, A.K.; Huberman, A.D.; Stafford, B.; et al. The classical complement cascade mediates CNS synapse elimination. *Cell* **2007**, *131*, 1164–1178. [CrossRef]
21. Chu, Y.; Jin, X.; Parada, I.; Pesic, A.; Stevens, B.; Barres, B.; Prince, D.A. Enhanced synaptic connectivity and epilepsy in C1q knockout mice. *Proc. Natl. Acad. Sci. USA* **2010**, *107*, 7975–7980. [CrossRef]
22. Sekar, A.; Bialas, A.R.; de Rivera, H.; Davis, A.; Hammond, T.R.; Kamitaki, N.; Tooley, K.; Presumey, J.; Baum, M.; Van Doren, V.; et al. Schizophrenia risk from complex variation of complement component 4. *Nature* **2016**, *530*, 177–183. [CrossRef]
23. Hong, S.; Beja-Glasser, V.F.; Nfonoyim, B.M.; Frouin, A.; Li, S.; Ramakrishnan, S.; Merry, K.M.; Shi, Q.; Rosenthal, A.; Barres, B.A.; et al. Complement and microglia mediate early synapse loss in Alzheimer mouse models. *Science* **2016**, *352*, 712–716. [CrossRef]
24. Hakobyan, S.; Luppe, S.; Evans, D.R.; Harding, K.; Loveless, S.; Robertson, N.P.; Morgan, B.P. Plasma complement biomarkers distinguish multiple sclerosis and neuromyelitis optica spectrum disorder. *Mult. Scler.* **2017**, *23*, 946–955. [CrossRef]

25. Morgan, B.P.; Gommerman, J.L.; Ramaglia, V. An "Outside-In" and "Inside-Out" Consideration of Complement in the Multiple Sclerosis Brain: Lessons from Development and Neurodegenerative Diseases. *Front. Cell Neurosci.* **2020**, *14*, 600656. [CrossRef]
26. Harold, D.; Abraham, R.; Hollingworth, P.; Sims, R.; Gerrish, A.; Hamshere, M.L.; Pahwa, J.S.; Moskvina, V.; Dowzell, K.; Williams, A.; et al. Genome-wide association study identifies variants at CLU and PICALM associated with Alzheimer's disease. *Nat. Genet.* **2009**, *41*, 1088–1093. [CrossRef]
27. Schifferli, J.A. Complement and immune complexes. *Res. Immunol.* **1996**, *147*, 109–110. [CrossRef]
28. Kavai, M. Immune complex clearance by complement receptor type 1 in SLE. *Autoimmun. Rev.* **2008**, *8*, 160–164. [CrossRef] [PubMed]
29. Gasque, P.; Chan, P.; Mauger, C.; Schouft, M.T.; Singhrao, S.; Dierich, M.P.; Morgan, B.P.; Fontaine, M. Identification and characterization of complement C3 receptors on human astrocytes. *J. Immunol.* **1996**, *156*, 2247–2255.
30. Hazrati, L.N.; Van Cauwenberghe, C.; Brooks, P.L.; Brouwers, N.; Ghani, M.; Sato, C.; Cruts, M.; Sleegers, K.; St George-Hyslop, P.; Van Broeckhoven, C.; et al. Genetic association of CR1 with Alzheimer's disease: A tentative disease mechanism. *Neurobiol. Aging* **2012**, *33*, 2949.e5–2959.e12. [CrossRef]
31. Fonseca, M.I.; Chu, S.; Pierce, A.L.; Brubaker, W.D.; Hauhart, R.E.; Mastroeni, D.; Clarke, E.V.; Rogers, J.; Atkinson, J.P.; Tenner, A.J. Analysis of the Putative Role of CR1 in Alzheimer's Disease: Genetic Association, Expression and Function. *PLoS ONE* **2016**, *11*, e0149792. [CrossRef]
32. Mancuso, R.; Van Den Daele, J.; Fattorelli, N.; Wolfs, L.; Balusu, S.; Burton, O.; Liston, A.; Sierksma, A.; Fourne, Y.; Poovathingal, S.; et al. Stem-cell-derived human microglia transplanted in mouse brain to study human disease. *Nat. Neurosci.* **2019**, *22*, 2111–2116. [CrossRef]
33. Haenseler, W.; Sansom, S.N.; Buchrieser, J.; Newey, S.E.; Moore, C.S.; Nicholls, F.J.; Chintawar, S.; Schnell, C.; Antel, J.P.; Allen, N.D.; et al. A Highly Efficient Human Pluripotent Stem Cell Microglia Model Displays a Neuronal-Co-culture-Specific Expression Profile and Inflammatory Response. *Stem Cell Rep.* **2017**, *8*, 1727–1742. [CrossRef]
34. Johansson, J.U.; Brubaker, W.D.; Javitz, H.; Bergen, A.W.; Nishita, D.; Trigunaite, A.; Crane, A.; Ceballos, J.; Mastroeni, D.; Tenner, A.J.; et al. Peripheral complement interactions with amyloid β peptide in Alzheimer's disease: Polymorphisms, structure, and function of complement receptor 1. *Alzheimers Dement.* **2018**, *14*, 1438–1449. [CrossRef]
35. Crehan, H.; Holton, P.; Wray, S.; Pocock, J.; Guerreiro, R.; Hardy, J. Complement receptor 1 (CR1) and Alzheimer's disease. *Immunobiology* **2012**, *217*, 244–250. [CrossRef]
36. Moulds, J.M.; Reveille, J.D.; Arnett, F.C. Structural polymorphisms of complement receptor 1 (CR1) in systemic lupus erythematosus (SLE) patients and normal controls of three ethnic groups. *Clin. Exp. Immunol.* **1996**, *105*, 302–305. [CrossRef]
37. Brouwers, N.; Van Cauwenberghe, C.; Engelborghs, S.; Lambert, J.C.; Bettens, K.; Le Bastard, N.; Pasquier, F.; Montoya, A.G.; Peeters, K.; Mattheijssens, M.; et al. Alzheimer risk associated with a copy number variation in the complement receptor 1 increasing C3b/C4b binding sites. *Mol. Psychiatry* **2012**, *17*, 223–233. [CrossRef]
38. Kucukkilic, E.; Brookes, K.; Barber, I.; Guetta-Baranes, T.; Consortium, A.; Morgan, K.; Hollox, E.J. Complement receptor 1 gene (CR1) intragenic duplication and risk of Alzheimer's disease. *Hum. Genet.* **2018**, *137*, 305–314. [CrossRef]
39. Krych-Goldberg, M.; Atkinson, J.P. Structure-function relationships of complement receptor type 1. *Immunol. Rev.* **2001**, *180*, 112–122. [CrossRef]
40. Mahmoudi, R.; Feldman, S.; Kisserli, A.; Duret, V.; Tabary, T.; Bertholon, L.A.; Badr, S.; Nonnonhou, V.; Cesar, A.; Neuraz, A.; et al. Inherited and Acquired Decrease in Complement Receptor 1 (CR1) Density on Red Blood Cells Associated with High Levels of Soluble CR1 in Alzheimer's Disease. *Int. J. Mol. Sci.* **2018**, *19*, 2175. [CrossRef]
41. Mahmoudi, R.; Kisserli, A.; Novella, J.L.; Donvito, B.; Drame, M.; Reveil, B.; Duret, V.; Jolly, D.; Pham, B.N.; Cohen, J.H. Alzheimer's disease is associated with low density of the long CR1 isoform. *Neurobiol. Aging* **2015**, *36*, 1766.e5–1766.e12. [CrossRef]
42. Liu, D.; Niu, Z.X. The structure, genetic polymorphisms, expression and biological functions of complement receptor type 1 (CR1/CD35). *Immunopharmacol. Immunotoxicol.* **2009**, *31*, 524–535. [CrossRef]
43. Keenan, B.T.; Shulman, J.M.; Chibnik, L.B.; Raj, T.; Tran, D.; Sabuncu, M.R.; Alzheimer's Disease Neuroimaging, I.; Allen, A.N.; Corneveaux, J.J.; Hardy, J.A.; et al. A coding variant in CR1 interacts with APOE-epsilon4 to influence cognitive decline. *Hum. Mol. Genet.* **2012**, *21*, 2377–2388. [CrossRef]
44. Jacquet, M.; Cioci, G.; Fouet, G.; Bally, I.; Thielens, N.M.; Gaboriaud, C.; Rossi, V. C1q and Mannose-Binding Lectin Interact with CR1 in the Same Region on CCP24-25 Modules. *Front. Immunol.* **2018**, *9*, 453. [CrossRef]
45. Van Cauwenberghe, C.; Bettens, K.; Engelborghs, S.; Vandenbulcke, M.; Van Dongen, J.; Vermeulen, S.; Vandenberghe, R.; De Deyn, P.P.; Van Broeckhoven, C.; Sleegers, K. Complement receptor 1 coding variant p.Ser1610Thr in Alzheimer's disease and related endophenotypes. *Neurobiol. Aging* **2013**, *34*, 2235.e1–2235.e6. [CrossRef]
46. Hollingworth, P.; Sweet, R.; Sims, R.; Harold, D.; Russo, G.; Abraham, R.; Stretton, A.; Jones, N.; Gerrish, A.; Chapman, J.; et al. Genome-wide association study of Alzheimer's disease with psychotic symptoms. *Mol. Psychiatry* **2012**, *17*, 1316–1327. [CrossRef] [PubMed]
47. Jansen, I.E.; Savage, J.E.; Watanabe, K.; Bryois, J.; Williams, D.M.; Steinberg, S.; Sealock, J.; Karlsson, I.K.; Hagg, S.; Athanasiu, L.; et al. Genome-wide meta-analysis identifies new loci and functional pathways influencing Alzheimer's disease risk. *Nat. Genet.* **2019**, *51*, 404–413. [CrossRef] [PubMed]

48. Naj, A.C.; Jun, G.; Beecham, G.W.; Wang, L.S.; Vardarajan, B.N.; Buros, J.; Gallins, P.J.; Buxbaum, J.D.; Jarvik, G.P.; Crane, P.K.; et al. Common variants at MS4A4/MS4A6E, CD2AP, CD33 and EPHA1 are associated with late-onset Alzheimer's disease. *Nat. Genet.* **2011**, *43*, 436–441. [CrossRef]
49. Zhang, Q.; Yu, J.T.; Zhu, Q.X.; Zhang, W.; Wu, Z.C.; Miao, D.; Tan, L. Complement receptor 1 polymorphisms and risk of late-onset Alzheimer's disease. *Brain Res.* **2010**, *1348*, 216–221. [CrossRef]
50. Jin, C.; Li, W.; Yuan, J.; Xu, W.; Cheng, Z. Association of the CR1 polymorphism with late-onset Alzheimer's disease in Chinese Han populations: A meta-analysis. *Neurosci. Lett.* **2012**, *527*, 46–49. [CrossRef]
51. Li, H.L.; Shi, S.S.; Guo, Q.H.; Ni, W.; Dong, Y.; Liu, Y.; Sun, Y.M.; Bei, W.; Lu, S.J.; Hong, Z.; et al. PICALM and CR1 variants are not associated with sporadic Alzheimer's disease in Chinese patients. *J. Alzheimers Dis.* **2011**, *25*, 111–117. [CrossRef]
52. Jenne, D.E.; Tschopp, J. Clusterin: The intriguing guises of a widely expressed glycoprotein. *Trends Biochem. Sci.* **1992**, *17*, 154–159. [CrossRef]
53. Humphreys, D.T.; Carver, J.A.; Easterbrook-Smith, S.B.; Wilson, M.R. Clusterin has chaperone-like activity similar to that of small heat shock proteins. *J. Biol. Chem.* **1999**, *274*, 6875–6881. [CrossRef] [PubMed]
54. Zhang, H.; Kim, J.K.; Edwards, C.A.; Xu, Z.; Taichman, R.; Wang, C.Y. Clusterin inhibits apoptosis by interacting with activated Bax. *Nat. Cell Biol.* **2005**, *7*, 909–915. [CrossRef]
55. Ammar, H.; Closset, J.L. Clusterin activates survival through the phosphatidylinositol 3-kinase/Akt pathway. *J. Biol. Chem.* **2008**, *283*, 12851–12861. [CrossRef]
56. Trougakos, I.P.; Lourda, M.; Antonelou, M.H.; Kletsas, D.; Gorgoulis, V.G.; Papassideri, I.S.; Zou, Y.; Margaritis, L.H.; Boothman, D.A.; Gonos, E.S. Intracellular clusterin inhibits mitochondrial apoptosis by suppressing p53-activating stress signals and stabilizing the cytosolic Ku70-Bax protein complex. *Clin. Cancer Res.* **2009**, *15*, 48–59. [CrossRef]
57. Tschopp, J.; French, L.E. Clusterin: Modulation of complement function. *Clin. Exp. Immunol.* **1994**, *97* (Suppl. 2), 11–14. [CrossRef]
58. McDonald, J.F.; Nelsestuen, G.L. Potent inhibition of terminal complement assembly by clusterin: Characterization of its impact on C9 polymerization. *Biochemistry* **1997**, *36*, 7464–7473. [CrossRef]
59. Pasinetti, G.M.; Johnson, S.A.; Oda, T.; Rozovsky, I.; Finch, C.E. Clusterin (SGP-2): A multifunctional glycoprotein with regional expression in astrocytes and neurons of the adult rat brain. *J. Comp. Neurol.* **1994**, *339*, 387–400. [CrossRef] [PubMed]
60. Gasque, P.; Fontaine, M.; Morgan, B.P. Complement expression in human brain. Biosynthesis of terminal pathway components and regulators in human glial cells and cell lines. *J. Immunol.* **1995**, *154*, 4726–4733.
61. Wiggins, A.K.; Shen, P.J.; Gundlach, A.L. Delayed, but prolonged increases in astrocytic clusterin (ApoJ) mRNA expression following acute cortical spreading depression in the rat: Evidence for a role of clusterin in ischemic tolerance. *Brain Res. Mol. Brain Res.* **2003**, *114*, 20–30. [CrossRef]
62. Calero, M.; Rostagno, A.; Matsubara, E.; Zlokovic, B.; Frangione, B.; Ghiso, J. Apolipoprotein J (clusterin) and Alzheimer's disease. *Microsc. Res. Tech.* **2000**, *50*, 305–315. [CrossRef]
63. Troakes, C.; Smyth, R.; Noor, F.; Maekawa, S.; Killick, R.; King, A.; Al-Sarraj, S. Clusterin expression is upregulated following acute head injury and localizes to astrocytes in old head injury. *Neuropathology* **2017**, *37*, 12–24. [CrossRef]
64. Anderson, A.J.; Najbauer, J.; Huang, W.; Young, W.; Robert, S. Upregulation of complement inhibitors in association with vulnerable cells following contusion-induced spinal cord injury. *J. Neurotrauma* **2005**, *22*, 382–397. [CrossRef]
65. Foster, E.M.; Dangla-Valls, A.; Lovestone, S.; Ribe, E.M.; Buckley, N.J. Clusterin in Alzheimer's Disease: Mechanisms, Genetics, and Lessons from Other Pathologies. *Front. Neurosci.* **2019**, *13*, 164. [CrossRef]
66. McGeer, P.L.; Kawamata, T.; Walker, D.G. Distribution of clusterin in Alzheimer brain tissue. *Brain Res.* **1992**, *579*, 337–341. [CrossRef]
67. Lidstrom, A.M.; Bogdanovic, N.; Hesse, C.; Volkman, I.; Davidsson, P.; Blennow, K. Clusterin (apolipoprotein J) protein levels are increased in hippocampus and in frontal cortex in Alzheimer's disease. *Exp. Neurol.* **1998**, *154*, 511–521. [CrossRef]
68. Bertrand, P.; Poirier, J.; Oda, T.; Finch, C.E.; Pasinetti, G.M. Association of apolipoprotein E genotype with brain levels of apolipoprotein E and apolipoprotein J (clusterin) in Alzheimer disease. *Brain Res. Mol. Brain Res.* **1995**, *33*, 174–178. [CrossRef]
69. Nilselid, A.M.; Davidsson, P.; Nagga, K.; Andreasen, N.; Fredman, P.; Blennow, K. Clusterin in cerebrospinal fluid: Analysis of carbohydrates and quantification of native and glycosylated forms. *Neurochem. Int.* **2006**, *48*, 718–728. [CrossRef]
70. Krance, S.H.; Wu, C.Y.; Zou, Y.; Mao, H.; Toufighi, S.; He, X.; Pakosh, M.; Swardfager, W. The complement cascade in Alzheimer's disease: A systematic review and meta-analysis. *Mol. Psychiatry* **2019**. [CrossRef]
71. Schrijvers, E.M.; Koudstaal, P.J.; Hofman, A.; Breteler, M.M. Plasma clusterin and the risk of Alzheimer disease. *JAMA* **2011**, *305*, 1322–1326. [CrossRef]
72. Jongbloed, W.; van Dijk, K.D.; Mulder, S.D.; van de Berg, W.D.; Blankenstein, M.A.; van der Flier, W.; Veerhuis, R. Clusterin Levels in Plasma Predict Cognitive Decline and Progression to Alzheimer's Disease. *J. Alzheimers Dis.* **2015**, *46*, 1103–1110. [CrossRef]
73. Hakobyan, S.; Harding, K.; Aiyaz, M.; Hye, A.; Dobson, R.; Baird, A.; Liu, B.; Harris, C.L.; Lovestone, S.; Morgan, B.P. Complement Biomarkers as Predictors of Disease Progression in Alzheimer's Disease. *J. Alzheimers Dis.* **2016**, *54*, 707–716. [CrossRef]
74. Ghiso, J.; Matsubara, E.; Koudinov, A.; Choi-Miura, N.H.; Tomita, M.; Wisniewski, T.; Frangione, B. The cerebrospinal-fluid soluble form of Alzheimer's amyloid β is complexed to SP-40,40 (apolipoprotein J), an inhibitor of the complement membrane-attack complex. *Biochem. J.* **1993**, *293 Pt 1*, 27–30. [CrossRef]

75. Zlokovic, B.V.; Martel, C.L.; Matsubara, E.; McComb, J.G.; Zheng, G.; McCluskey, R.T.; Frangione, B.; Ghiso, J. Glycoprotein 330/megalin: Probable role in receptor-mediated transport of apolipoprotein J alone and in a complex with Alzheimer disease amyloid β at the blood-brain and blood-cerebrospinal fluid barriers. *Proc. Natl. Acad. Sci. USA* **1996**, *93*, 4229–4234. [CrossRef]
76. Narayan, P.; Meehan, S.; Carver, J.A.; Wilson, M.R.; Dobson, C.M.; Klenerman, D. Amyloid-β oligomers are sequestered by both intracellular and extracellular chaperones. *Biochemistry* **2012**, *51*, 9270–9276. [CrossRef]
77. DeMattos, R.B.; O'Dell, M.A.; Parsadanian, M.; Taylor, J.W.; Harmony, J.A.; Bales, K.R.; Paul, S.M.; Aronow, B.J.; Holtzman, D.M. Clusterin promotes amyloid plaque formation and is critical for neuritic toxicity in a mouse model of Alzheimer's disease. *Proc. Natl. Acad. Sci. USA* **2002**, *99*, 10843–10848. [CrossRef]
78. De Mattos, R.B.; Cirrito, J.R.; Parsadanian, M.; May, P.C.; O'Dell, M.A.; Taylor, J.W.; Harmony, J.A.; Aronow, B.J.; Bales, K.R.; Paul, S.M.; et al. ApoE and clusterin cooperatively suppress Abeta levels and deposition: Evidence that ApoE regulates extracellular Abeta metabolism in vivo. *Neuron* **2004**, *41*, 193–202. [CrossRef]
79. Jackson, R.J.; Rose, J.; Tulloch, J.; Henstridge, C.; Smith, C.; Spires-Jones, T.L. Clusterin accumulates in synapses in Alzheimer's disease and is increased in apolipoprotein E4 carriers. *Brain Commun.* **2019**, *1*, fcz003. [CrossRef]
80. Bettens, K.; Vermeulen, S.; Van Cauwenberghe, C.; Heeman, B.; Asselbergh, B.; Robberecht, C.; Engelborghs, S.; Vandenbulcke, M.; Vandenberghe, R.; De Deyn, P.P.; et al. Reduced secreted clusterin as a mechanism for Alzheimer-associated CLU mutations. *Mol. Neurodegener.* **2015**, *10*, 30. [CrossRef]
81. Schurmann, B.; Wiese, B.; Bickel, H.; Weyerer, S.; Riedel-Heller, S.G.; Pentzek, M.; Bachmann, C.; Williams, J.; van den Bussche, H.; Maier, W.; et al. Association of the Alzheimer's disease clusterin risk allele with plasma clusterin concentration. *J. Alzheimers Dis.* **2011**, *25*, 421–424. [CrossRef]
82. Xing, Y.Y.; Yu, J.T.; Cui, W.Z.; Zhong, X.L.; Wu, Z.C.; Zhang, Q.; Tan, L. Blood clusterin levels, rs9331888 polymorphism, and the risk of Alzheimer's disease. *J. Alzheimers Dis.* **2012**, *29*, 515–519. [CrossRef] [PubMed]
83. Cai, R.; Han, J.; Sun, J.; Huang, R.; Tian, S.; Shen, Y.; Dong, X.; Xia, W.; Wang, S. Plasma Clusterin and the CLU Gene rs11136000 Variant Are Associated with Mild Cognitive Impairment in Type 2 Diabetic Patients. *Front. Aging Neurosci.* **2016**, *8*, 179. [CrossRef]
84. Carrasquillo, M.M.; Crook, J.E.; Pedraza, O.; Thomas, C.S.; Pankratz, V.S.; Allen, M.; Nguyen, T.; Malphrus, K.G.; Ma, L.; Bisceglio, G.D.; et al. Late-onset Alzheimer's risk variants in memory decline, incident mild cognitive impairment, and Alzheimer's disease. *Neurobiol. Aging* **2015**, *36*, 60–67. [CrossRef]
85. Lin, Y.L.; Chen, S.Y.; Lai, L.C.; Chen, J.H.; Yang, S.Y.; Huang, Y.L.; Chen, T.F.; Sun, Y.; Wen, L.L.; Yip, P.K.; et al. Genetic polymorphisms of clusterin gene are associated with a decreased risk of Alzheimer's disease. *Eur. J. Epidemiol.* **2012**, *27*, 73–75. [CrossRef] [PubMed]
86. Sampedro, F.; Marin-Lahoz, J.; Martinez-Horta, S.; Perez-Gonzalez, R.; Pagonabarraga, J.; Kulisevsky, J. CLU rs11136000 promotes early cognitive decline in Parkinson's disease. *Mov. Disord.* **2020**, *35*, 508–513. [CrossRef]
87. Thambisetty, M.; Simmons, A.; Velayudhan, L.; Hye, A.; Campbell, J.; Zhang, Y.; Wahlund, L.O.; Westman, E.; Kinsey, A.; Guntert, A.; et al. Association of plasma clusterin concentration with severity, pathology, and progression in Alzheimer disease. *Arch. Gen. Psychiatry* **2010**, *67*, 739–748. [CrossRef]
88. Szymanski, M.; Wang, R.; Bassett, S.S.; Avramopoulos, D. Alzheimer's risk variants in the clusterin gene are associated with alternative splicing. *Transl. Psychiatry* **2011**, *1*, e18. [CrossRef]
89. Uddin, M.S.; Kabir, M.T.; Begum, M.M.; Islam, M.S.; Behl, T.; Ashraf, G.M. Exploring the Role of *CLU* in the Pathogenesis of Alzheimer's Disease. *Neurotox. Res.* **2020**, *39*, 2108–2119. [CrossRef] [PubMed]
90. Nguyen, V.C.; Tosi, M.; Gross, M.S.; Cohen-Haguenauer, O.; Jegou-Foubert, C.; de Tand, M.F.; Meo, T.; Frezal, J. Assignment of the complement serine protease genes C1r and C1s to chromosome 12 region 12p13. *Hum. Genet.* **1988**, *78*, 363–368. [CrossRef] [PubMed]
91. Zhang, Y.; Sloan, S.A.; Clarke, L.E.; Caneda, C.; Plaza, C.A.; Blumenthal, P.D.; Vogel, H.; Steinberg, G.K.; Edwards, M.S.; Li, G.; et al. Purification and Characterization of Progenitor and Mature Human Astrocytes Reveals Transcriptional and Functional Differences with Mouse. *Neuron* **2016**, *89*, 37–53. [CrossRef] [PubMed]
92. Khoonsari, P.E.; Haggmark, A.; Lonnberg, M.; Mikus, M.; Kilander, L.; Lannfelt, L.; Bergquist, J.; Ingelsson, M.; Nilsson, P.; Kultima, K.; et al. Analysis of the Cerebrospinal Fluid Proteome in Alzheimer's Disease. *PLoS ONE* **2016**, *11*, e0150672. [CrossRef]
93. Raybould, R.; Sims, R. Searching the Dark Genome for Alzheimer's Disease Risk Variants. *Brain Sci.* **2021**, *11*, 332. [CrossRef]
94. Ebbert, M.T.W.; Jensen, T.D.; Jansen-West, K.; Sens, J.P.; Reddy, J.S.; Ridge, P.G.; Kauwe, J.S.K.; Belzil, V.; Pregent, L.; Carrasquillo, M.M.; et al. Systematic analysis of dark and camouflaged genes reveals disease-relevant genes hiding in plain sight. *Genome Biol.* **2019**, *20*, 97. [CrossRef]
95. Carpanini, S.M.; Harwood, J.C.; Baker, E.; Torvell, M.; The GERARD Consortium; Sims, R.; Williams, J.; Morgan, B.P. The Impact of Complement Genes on the Risk of Late-Onset Alzheimer's Disease. *Genes* **2021**, *12*, 443. [CrossRef]
96. Carpanini, S.M.; Torvell, M.; Morgan, B.P. Therapeutic Inhibition of the Complement System in Diseases of the Central Nervous System. *Front. Immunol.* **2019**, *10*, 362. [CrossRef]

Review

Advances with Long Non-Coding RNAs in Alzheimer's Disease as Peripheral Biomarker

Maria Garofalo [1,2,†], Cecilia Pandini [1,2,†], Daisy Sproviero [1], Orietta Pansarasa [1], Cristina Cereda [1] and Stella Gagliardi [1,*]

1. Genomic and Post-Genomic Unit, IRCCS Mondino Foundation, 27100 Pavia, Italy; maria.garofalo@mondino.it (M.G.); cecilia.pandini@mondino.it (C.P.); daisy.sproviero@mondino.it (D.S.); orietta.pansarasa@mondino.it (O.P.); cristina.cereda@mondino.it (C.C.)
2. Department of Biology and Biotechnology "L. Spallanzani", University of Pavia, 27100 Pavia, Italy
* Correspondence: stella.gagliardi@mondino.it; Tel.: +39-038-238-0248
† These authors contributed equally.

Abstract: One of the most compelling needs in the study of Alzheimer's disease (AD) is the characterization of cognitive decline peripheral biomarkers. In this context, the theme of altered RNA processing has emerged as a contributing factor to AD. In particular, the significant role of long non-coding RNAs (lncRNAs) associated to AD is opening new perspectives in AD research. This class of RNAs may offer numerous starting points for new investigations about pathogenic mechanisms and, in particular, about peripheral biomarkers. Indeed, altered lncRNA signatures are emerging as potential diagnostic biomarkers. In this review, we have collected and fully explored all the presented data about lncRNAs and AD in the peripheral system to offer an overview about this class of non-coding RNAs and their possible role in AD.

Keywords: long non-coding RNA; Alzheimer's disease; biomarkers; peripheral system

1. RNA Metabolism in Alzheimer's Disease

Alzheimer's disease (AD) is a progressive neurodegenerative disorder that leads to intellectual functions' impairment. AD is the most common type of dementia in aging populations causing neuropathology in specific brain regions, including hippocampus, amygdala, and frontal and temporal cortices. Complex multifactorial interactions among genetic, epigenetic, and environmental components contribute to AD onset. Although much emphasis has been placed on the role of protein aggregates (Aβ plaques and tau tangles) in AD, recent multiple lines of evidence converge on altered RNA metabolism as a contributing factor in the pathogenesis of this disorder. In particular, non-coding RNAs' role is emerging as involved in pathogenesis, diagnosis and therapy of AD. For instance, many microRNAs (miRNAs) have been identified as key elements for the regulation of memory process and cognitive functions lost in AD [1]. They can act through the regulation of activity-mediated protein synthesis at the synaptic level [2], the regulation of Aβ production [3,4] and tau phosphorylation [3]. Circular RNAs (circRNAs), a type of single-stranded RNA which forms a covalently closed continuous loop, can act as a miRNA "sponge" to quench normal miRNA functions [5]. This mechanism has been found also in AD, where the altered circRNA ciRS-7 sponging activity for miRNA-7 leads to the lack of essential proteins for the clearance of amyloid peptides in AD brain [6]. Moreover, mounting evidence shows that long non-coding RNAs (lncRNAs) are aberrantly expressed in AD progression and participate in the regulation of Aβ peptide [7,8] tau [9], inflammation and cell death [10,11].

2. Long Non-Coding RNAs

LncRNAs are defined as non-coding RNA molecules longer than 200 nucleotides. Most of them are transcribed by RNA polymerase II and are often post transcriptionally modified by splicing, 5′ 7-methylguanosine capping and a 3′ polyadenylation; however, they lack coding capacity [12]. Human GENCODE suggests that the human genome contains more than 16,000 lncRNA genes, but other estimates exceed 100,000 human lncRNAs [13]. Despite not being translated into proteins, lncRNAs are functional molecules with high heterogeneity and functional versatility that relies on their ability as long RNA molecules to conform to different structures and molecular interactions. Indeed, lncRNAs can regulate, among other things, transcriptional regulation in cis or trans, organization of nuclear domains, and regulation of proteins or RNA molecules, affecting numerous biological and pathological processes [14].

3. lncRNAs in AD Peripheral System

3.1. Blood

The discovery of peripheral biomarkers for neurodegenerative disease, such as AD, is needed. LncRNAs may be a noninvasive target to confirm AD diagnosis and they can also be used as prognostic biomarkers.

Different papers have investigated lncRNAs in blood for AD patients. Kurt and collaborators [15] have investigated lncRNAs' expression difference between AD patients and controls in peripheral blood mononuclear cell (PBMC) by microarray analysis. Their data showed that 34 lncRNAs have been found deregulated, in particular the most altered lncRNA is an antisense transcript named TTC39C-AS1. This antisense is interesting since its sense gene, *TTC39C*, is involved in neurogenic atrophy [16]. Next, another highly deregulated lncRNA was LOC401557 that is an uncharacterized lncRNA very abundant in the brain tissue [17]. Gene deregulation generally implicates changes in gene expression altering cell homeostasis, and its understanding may provide new insights into the mechanisms involved in human diseases [18]. In general, different pathways in which lncRNA may have a role have been identified, such as amyloidogenic and mTOR pathways. For both, a deregulation of lncRNAs occurs as LINC01503 and LINC01420 are altered in PBMCs and also in brain [19,20].

We previously demonstrated deregulated lncRNAs in PBMCs from AD patients by RNA-seq. We compared the lncRNA profile of AD patients with two other neurodegenarative diseases, Parkinson's disease and amyotrophic lateral sclerosis [21]. The data showed that CH507-513H4.4, CH507-513H4.6, CH507-513H4.3 lncRNAs are deregulated in AD PBMC compared to controls. They are novel transcripts, similar to YY1 Associated Myogenesis RNA 1 (YAM1), and they are reported as AD associated in the LncRNADisease v2.0 Database [22]. These lncRNAs were specific for AD-in fact, no deregulation was found in the other diseases. Moreover, lncRNA pathway analysis was performed using the LncPath R package that showed an involvement of Mapk signaling, cytokine receptor interaction, chemokine signaling, natural killer cell mediated cytotoxicity and regulation of actin cytoskeleton.

3.2. Plasma

Two main plasma lncRNAs have been proposed as possible AD biomarkers: BACE1-AS and 51A [23].

51A is the antisense transcript of *SORL1* gene that was described as associated to AD for the first time in 2004, but its role is not clear [24]. SORL1 is involved in APP processing and trafficking. It may bind newly made Aβ in the neuron and steers it toward lysosomes, where it is degraded [25,26]. Besides this, SORL1 as an ApoE receptor is likely to participate in the lipid metabolism of AD genesis [27].

SORL1-AS (51A) expression leads to Aβ-42 accumulation, and it has been found to be increased in plasma and brain of AD patients compared to controls [28]. Clinical correlation showed that lncRNA 51A was negatively correlated with the Mini-Mental State Examination (MMSE) scores in AD patients.

About LncRNA BACE1-AS, its plasma level in AD patients was significantly higher compared to controls [29], while there was no correlation with MMSE scores. On the other hand, it has recently been demonstrated that lncRNA BACE1-AS may discriminate between full AD and controls but also between pre-AD and controls, suggesting that lncRNAs could be a predictive biomarker [30]. BACE1-AS regulates BACE1 mRNA and protein expression and may also increase BACE1 stability [8]. In fact, when BACE1-AS is silenced, the activity of BACE1 mRNA is attenuated and the production of Aβ-42 oligomers is reduced [31].

3.3. Extracellular Vesicles (EVs)

The presence of lncRNAs is also observed in extracellular vesicles (EVs). EVs are heterogenous lipid bound vesicles that are released and circulate in the extra-cellular space [32]. The two main subtypes of EVs are microvesicles (MVs), mostly derived from plasma membrane and 100–500 nm in diameter, and exosomes, generated through the classical endosome-multivesicular body (MVB) pathway and 30–150 nm in diameter [33]. The International Society for Extracellular Vesicles (ISEV) has updated EVs' nomenclature, defining as small EVs (SEVs) particles that are <100 nm or <200 nm and large EVs (LEVs) those that are >200 nm [33].

LncRNAs have mostly been observed packaged into SEVs [34,35]. SEVs can be released by practically all eukaryotic cells [36]. We found two studies concerning lncRNAs in AD in SEVs derived from plasma and cerebrospinal fluid (CSF).

BACE1-AS transcript was measured in plasma-derived SEVs from 72 AD and 62 controls. The level of this transcript was different in the two groups, being significantly higher in AD patients [37]. This result is in contrast with a previous study, that analyzed a smaller cohort of subjects, where the level of BACE1-AS remained unchanged in AD plasma SEVs [30].

BACE1-AS is able to influence the expression of Aβ and is described in AD pathogenesis [38]. Given the need of improving accuracy of AD diagnosis, Wang and collaborators tried to link pathological changes in the brain and the altered expression of BACE1-AS. However, they found no correlation between this lncRNA and Magnetic Resonance Imaging (MRI) data. Nevertheless, they also performed a receiver operating characteristic (ROC) curve analysis, which is a graphical approach for comparing the relative performance of different classifiers and to determine whether a classifier performs better than random guessing [39]. They demonstrated that when exosomal BACE1-AS levels are combined with the volume and thickness of the right entorhinal cortex, specificity and sensitivity were at high percentage, making these parameters potential biomarkers of AD [37].

The expression of two lncRNAs, RP11-462G22.1 and PCA3, was also evaluated in CSF-derived SEVs from AD patients [40]. These two transcripts were found to be associated with Parkinson's disease (PD). These lncRNAs may not represent the perfect biomarkers for discriminating AD and PD, due to the fact that they are deregulated in both conditions, but they could rather be used as indicative molecules for neurodegeneration. RP11-462G22.1, instead, was found to be highly expressed in AD and PD. It is a muscular dystrophy-associated lncRNA that was predicted to be the target of 21 microRNAs, making it a potential competing endogenous RNA (ceRNA) [41]. PCA3, another lncRNA up-regulated in CSF-derived SEVs from AD patients, may be targeted by 14 microRNAs [42]. PCA3's biological function in neurodegenerative disorders is still unknown.

So far, the study of lncRNAs in EVs from AD patients is not sufficient for providing informative evidence of their role in the pathogenesis of this disease. Nor has a relevant screening of these molecules been published in order to highlight reliable biomarkers that could be used in the diagnosis or prognosis of AD.

3.4. Cerebrospinal Fluid (CSF)

The most instructive fluid in biomarker detection for neurodegeneration is cerebrospinal Fluid (CSF) [43]. Thus, we explored literature in order to highlight the most promising lncRNAs studied in CSF of AD patients.

MALAT1, a long intergenic non-coding RNA, regulates synaptogenesis and, in fact, its expression is widely observed in neurons [44]. It may be used as a diagnostic biomarker of AD in CSF, where it was found down-regulated [45]. The role of MALAT1 was initially described in AD models where the expression of the transcript was both up and down-regulated [11]. In this study, enhanced neuron apoptosis, repressed neurite outgrowth and elevated inflammation-related molecules were observed where MALAT1 levels were lower. Moreover, they found miR-125b, which induces the processes listed above, to be negatively affected by MALAT1. Thus, low levels of lncRNA MALAT1 promote miR-125b enrichment, which in turn increases prostaglandin-endoperoxide synthase 2 (*PTGS2*) and cyclin-dependent kinase 5 (*CDK5*) expression levels and decreased forkhead box Q1 (*FQXQ1*). Interestingly, the intercorrelation of MALAT1 and miR-125b with *FOXQ1*, *PTGS2* and *CDK5* was also confirmed in CSF of AD patients [46]. In addition to functional characterization, this lncRNA–miRNA axis in CSF was also used for predicting Mini-Mental State Examination (MMSE) score decline at 1 year, 2 years and 3 years in AD patients.

Glial cell-derived neurotrophic factor (GDNF) is involved in neurite branching and synaptic plasticity [47]. In CSF of AD patients, GDNF mRNA is highly up-regulated [48]. The identification of a cis-antisense non-coding RNA to GDNF (GDNF-AS1 or GDNFOS) and its dependence to GDNF expression led Airavaara and collaborators to speculate that GDNF-AS1 may also be involved in synaptic plasticity and that further studies are needed to demonstrate the implication of this lncRNA in AD pathogenesis [47].

Long non-coding RNA activated by TGF-beta (lncRNA-ATB), firstly identified in 2014 [49], is abnormally expressed in central nervous system cancers [50]. Its expression was also altered in CSF of AD patients, where it was highly increased [51]. For this reason, deregulation of lncRNA-ATB may be used as a hallmark of disease rather than a specific biomarker. Moreover, in a recent study adult malignant brain tumors and AD were found to share some environmental risks [52]. LncRNA-ATB is indeed up-regulated both in AD patients and in glioma tumors.

To study the effect of lncRNA-ATB up-regulation, Wang and collaborators used PC12 cells and discovered that miR-200 is negatively affected by this lncRNA. MiR-200 in turn inversely regulates makorin ring finger protein 3 (MKRN3 or ZNF127), which is a 3-ubiquitin ligase potentially affecting gene expression and targeted protein degradation [47]. The inhibition of miR-200 mediated by lncRNA-ATB overexpression aggravated PC12 cells injury induced through Aβ25-35 [51]. However, the role of ZNF127 in neurodegeneration remains unclear. Altogether, these results highlight the relevance of the lncRNA–ATB/miR-200 axis in AD (Table 1).

Table 1. Deregulated lncRNA in peripheral tissue of AD patients.

Deregulated lncRNA in AD	Trend	Source	Reference	Tissue Expression
TTC39C-AS1	up-regulated	Blood	[1–6]	adrenal; brain; breast; lymphnode; testes; thyroid
LOC401557	up-regulated	Blood	[17]	adipose; adrenal; brain; breast; colon; foreskin; heart; HLF; kidney; liver; lung; lymphnode; ovary; placenta; prostate; skeletal muscle; testes; thyroid; WBC
CH507-513H4.4	up-regulated	Blood	[21]	/
CH507-513H4.6	up-regulated	Blood	[21]	/
CH507-513H4.3	up-regulated	Blood	[21]	/
SORL1-AS (51A)	up-regulated	Plasma	[28]	/
BACE1-AS	up-regulated	Plasma	[30]	brain; ovary; testes; thyroid
BACE1-AS	up-regulated	Plasma SEVs	[37]	brain; ovary; testes; thyroid
RP11-462G22.1	up-regulated	CSF SEVs	[40]	adipose; adrenal; brain; breast; colon; foreskin; heart; HLF; kidney; liver; lung; lymphnode; ovary; placenta; prostate; skeletal muscle; testes; thyroid; WBC
PCA3	up-regulated	CSF SEVs	[40]	brain; HLF; kidney; lymphnode; ovary; prostate; testes
MALAT1	down-regulated	CSF	[45]	adipose; brain; breast; lymphnode; prostate; testes; thyroid
lncRNA-ATB	up-regulated	CSF	[51]	adrenal; brain; breast; heart; HLF; liver; ovary; testes; thyroid

Deregulated lncRNAs in AD patients are reported together with their trend (up or down-regulated), their biofluid source and the corresponding literature reference. Using NONCODE database (www.noncode.org), human tissue where relative lncRNA was detected is reported. Human lung fibroblast (HLF); White blood cells (WBC).

4. Conclusions

In this review, we have explored literature classifying the current knowledge about lncRNAs in peripheral tissue of Alzheimer's disease patients. We found that several non-coding transcripts have been identified as potential biomarkers of this disease. Moreover, some studies have also highlighted the need to characterize the functional role of these molecules in the pathogenesis of Alzheimer's. In particular, the up-regulation of BACE1-AS in different tissues from AD patients appears to be a promising lncRNA in the study of AD due to its involvement in β-secretase regulation.

In conclusion, research concerning lncRNAs in neurodegenerative pathogenesis needs to be implemented in the future, covering both their potential as biomarkers and as therapeutic targets.

Author Contributions: S.G. and C.C.: idealization, intellectual input; M.G., C.P. and S.G.: literature search and writing the initial version of the manuscript; O.P. and D.S.: manuscript editing; S.G., C.C.: manuscript editing and supervision. All authors have read and agreed to the published version of the manuscript.

Funding: EuroNanoMed III JTC 2018 and Italian Ministry of Health; Fondazione Cariplo 2017 (Extracellular vesicles in the pathogenesis of Frontotemporal Dementia 2017-0747; Association between frailty trajectories and biological markers of aging 2017-0557).

Conflicts of Interest: The authors declare no conflict of interest.

References

1. Angelucci, F.; Cechova, K.; Valis, M.; Kuca, K.; Zhang, B.; Hort, J. MicroRNAs in Alzheimer's disease: Diagnostic markers or therapeutic agents? *Front. Pharmacol.* **2019**, *10*, 1–9. [CrossRef]
2. Ramakrishna, S.; Muddashetty, R.S. Emerging Role of microRNAs in Dementia. *J. Mol. Biol.* **2019**, *431*, 1743–1762. [CrossRef]
3. Hu, Y.K.; Wang, X.; Li, L.; Du, Y.H.; Ye, H.T.; Li, C.Y. MicroRNA-98 induces an Alzheimer's disease-like disturbance by targeting insulin-like growth factor 1. *Neurosci. Bull.* **2013**, *29*, 745–751. [CrossRef]
4. Fang, M.; Wang, J.; Zhang, X.; Geng, Y.; Hu, Z.; Rudd, J.A.; Ling, S.; Chen, W.; Han, S. The miR-124 regulates the expression of BACE1/β-secretase correlated with cell death in Alzheimer's disease. *Toxicol. Lett.* **2012**, *209*, 94–105. [CrossRef]
5. Hansen, T.B.; Jensen, T.I.; Clausen, B.H.; Bramsen, J.B.; Finsen, B.; Damgaard, C.K.; Kjems, J. Natural RNA circles function as efficient microRNA sponges. *Nature* **2013**, *495*, 384–388. [CrossRef] [PubMed]
6. Akhter, R. Circular RNA and Alzheimer's Disease. In *Circular RNAs: Biogenesis and Functions*; Xiao, J., Ed.; Springer: Singapore, 2018; pp. 239–243. ISBN 978-981-13-1426-1.
7. Massone, S.; Ciarlo, E.; Vella, S.; Nizzari, M.; Florio, T.; Russo, C.; Cancedda, R.; Pagano, A. NDM29, a RNA polymerase III-dependent non coding RNA, promotes amyloidogenic processing of APP and amyloid β secretion. *Biochim. Biophys. Acta Mol. Cell Res.* **2012**, *1823*, 1170–1177. [CrossRef]
8. Faghihi, M.A.; Modarresi, F.; Khalil, A.M.; Wood, D.E.; Sahagan, B.G.; Morgan, T.E.; Finch, C.E.; St. Laurent, G.; Kenny, P.J.; Wahlestedt, C. Expression of a noncoding RNA is elevated in Alzheimer's disease and drives rapid feed-forward regulation of β-secretase. *Nat. Med.* **2008**, *14*, 723–730. [CrossRef] [PubMed]
9. Ke, S.; Yang, Z.; Yang, F.; Wang, X.; Tan, J.; Liao, B. Long noncoding RNA NEAT1 aggravates Aβ-induced neuronal damage by targeting miR-107 in Alzheimer's disease. *Yonsei Med. J.* **2019**, *60*, 640–650. [CrossRef] [PubMed]
10. Wang, H.; Lu, B.; Chen, J. Biochemical and Biophysical Research Communications Knockdown of lncRNA SNHG1 attenuated A b 25-35 -inudced neuronal injury via regulating KREMEN1 by acting as a ceRNA of miR-137 in neuronal cells. *Biochem. Biophys. Res. Commun.* **2019**, *518*, 438–444. [CrossRef] [PubMed]
11. Ma, P.; Li, Y.; Zhang, W.; Fang, F.; Sun, J.; Liu, M.; Li, K.; Dong, L. Long Non-coding RNA MALAT1 Inhibits Neuron Apoptosis and Neuroinflammation While Stimulates Neurite Outgrowth and Its Correlation With MiR-125b Mediates PTGS2, CDK5 and FOXQ1 in Alzheimer's Disease. *Curr. Alzheimer Res.* **2019**, *16*, 596–612. [CrossRef]
12. Hon, C.C.; Ramilowski, J.A.; Harshbarger, J.; Bertin, N.; Rackham, O.J.L.; Gough, J.; Denisenko, E.; Schmeier, S.; Poulsen, T.M.; Severin, J.; et al. An atlas of human long non-coding RNAs with accurate 5′ ends. *Nature* **2017**, *543*, 199–204. [CrossRef] [PubMed]
13. Fang, S.; Zhang, L.; Guo, J.; Niu, Y.; Wu, Y.; Li, H.; Zhao, L.; Li, X.; Teng, X.; Sun, X.; et al. NONCODEV5: A comprehensive annotation database for long non-coding RNAs. *Nucleic Acids Res.* **2018**, *46*, D308–D314. [CrossRef] [PubMed]
14. Statello, L.; Guo, C.J.; Chen, L.L.; Huarte, M. Gene regulation by long non-coding RNAs and its biological functions. *Nat. Rev. Mol. Cell Biol.* **2021**, *22*, 96–118. [CrossRef] [PubMed]
15. Kurt, S.; Tomatir, A.G.; Tokgun, P.E.; Oncel, C. Altered Expression of Long Non-coding RNAs in Peripheral Blood Mononuclear Cells of Patients with Alzheimer's Disease. *Mol. Neurobiol.* **2020**, *57*, 5352–5361. [CrossRef]
16. Hayes, C.S.; Labuzan, S.A.; Menke, J.A.; Haddock, A.N.; Waddell, D.S. Ttc39c is upregulated during skeletal muscle atrophy and modulates ERK1/2 MAP kinase and hedgehog signaling. *J. Cell. Physiol.* **2019**, *234*, 23807–23824. [CrossRef]
17. Hill, S.E.; Donegan, R.K.; Nguyen, E.; Desai, T.M.; Lieberman, R.L. Molecular details of olfactomedin domains provide pathway to structure-function studies. *PLoS ONE* **2015**, *10*, 1–17. [CrossRef]
18. Lee, T.I.; Young, R.A. Transcriptional regulation and its misregulation in disease. *Cell* **2013**, *152*, 1237–1251. [CrossRef]
19. Zhou, X.; Xu, J. Identification of Alzheimer's disease-associated long noncoding RNAs. *Neurobiol. Aging* **2015**, *36*, 2925–2931. [CrossRef]
20. Mandas, A.; Abete, C.; Putzu, P.F.; La Colla, P.; Dess, S.; Pani, A. Changes in cholesterol metabolism-related gene expression in peripheral blood mononuclear cells from Alzheimer patients. *Lipids Health Dis.* **2012**, *11*, 1–8. [CrossRef]
21. Garofalo, M.; Pandini, C.; Bordoni, M.; Pansarasa, O.; Rey, F.; Costa, A.; Minafra, B.; Diamanti, L.; Zucca, S.; Carelli, S.; et al. Alzheimer's, parkinson's disease and amyotrophic lateral sclerosis gene expression patterns divergence reveals different grade of RNA metabolism involvement. *Int. J. Mol. Sci.* **2020**, *21*, 9500. [CrossRef]
22. Bao, Z.; Yang, Z.; Huang, Z.; Zhou, Y.; Cui, Q.; Dong, D. LncRNADisease 2.0: An updated database of long non-coding RNA-associated diseases. *Nucleic Acids Res.* **2019**, *47*, D1034–D1037. [CrossRef]
23. Wiinow, T.E.; Andersen, O.M. Sorting receptor SORLA—A trafficking path to avoid Alzheimer disease. *J. Cell Sci.* **2013**, *126*, 2751–2760. [CrossRef]
24. Scherzer, C.; Offe, K.; Lah, J.J. Loss of Apolipoprotein E Receptor. *Arch. Neurol.* **2004**, *61*, 1200–1205. [CrossRef] [PubMed]
25. Andersen, O.M.; Reiche, J.; Schmidt, V.; Gotthardt, M.; Spoelgen, R.; Behlke, J.; Von Arnim, C.A.F.; Breiderhoff, T.; Jansen, P.; Wu, X.; et al. Neuronal sorting protein-related receptor sorLA/LR11 regulates processing of the amyloid precursor protein. *Proc. Natl. Acad. Sci. USA* **2005**, *102*, 13461–13466. [CrossRef]
26. Caglayan, S.; Takagi-Niidome, S.; Liao, F.; Carlo, A.S.; Schmidt, V.; Burgert, T.; Kitago, Y.; Füchtbauer, E.M.; Füchtbauer, A.; Holtzman, D.M.; et al. Lysosomal sorting of amyloid-β by the SORLA receptor is impaired by a familial Alzheimer's disease mutation. *Sci. Transl. Med.* **2014**, *6*. [CrossRef]
27. Bu, G. Apolipoprotein E and its receptors in Alzheimer's disease: Pathways, pathogenesis and therapy. *Nat. Rev. Neurosci.* **2009**, *10*, 333–344. [CrossRef] [PubMed]

28. Deng, Y.; Xiao, L.; Li, W.; Tian, M.; Feng, X.; Feng, H.; Hou, D. Plasma long noncoding RNA 51A as a stable biomarker of Alzheimer's disease. *Int. J. Clin. Exp. Pathol.* **2017**, *10*, 4694–4699.
29. Feng, L.; Liao, Y.T.; He, J.C.; Xie, C.L.; Chen, S.Y.; Fan, H.H.; Su, Z.P.; Wang, Z. Plasma long non-coding RNA BACE1 as a novel biomarker for diagnosis of Alzheimer disease. *BMC Neurol.* **2018**, *18*, 1–8. [CrossRef]
30. Fotuhi, S.N.; Khalaj-kondori, M.; Feizi, M.A.H.; Talebi, M. Long Non-coding RNA BACE1-AS May Serve as an Alzheimer's Disease. *J. Mol. Neurosci.* **2019**, *69*, 351–359. [CrossRef]
31. Liu, T.; Huang, Y.; Chen, J.; Chi, H.; Yu, Z.; Wang, J.; Chen, C. Attenuated ability of BACE1 to cleave the amyloid precursor protein via silencing long noncoding RNA BACE1-AS expression. *Mol. Med. Rep.* **2014**, *10*, 1275–1281. [CrossRef]
32. Yáñez-Mó, M.; Siljander, P.R.M.; Andreu, Z.; Zavec, A.B.; Borràs, F.E.; Buzas, E.I.; Buzas, K.; Casal, E.; Cappello, F.; Carvalho, J.; et al. Biological properties of extracellular vesicles and their physiological functions. *J. Extracell. Vesicles* **2015**, *4*, 1–60. [CrossRef] [PubMed]
33. Théry, C.; Witwer, K.W.; Aikawa, E.; Alcaraz, M.J.; Anderson, J.D.; Andriantsitohaina, R.; Antoniou, A.; Arab, T.; Archer, F.; Atkin-Smith, G.K.; et al. Minimal information for studies of extracellular vesicles 2018 (MISEV2018): A position statement of the International Society for Extracellular Vesicles and update of the MISEV2014 guidelines. *J. Extracell. Vesicles* **2018**, *7*. [CrossRef] [PubMed]
34. Hewson, C.; Capraro, D.; Burdach, J.; Whitaker, N.; Morris, K.V. Extracellular vesicle associated long non-coding RNAs functionally enhance cell viability. *Non-Coding RNA Res.* **2016**, *1*, 3–11. [CrossRef]
35. Xu, Y.Z.; Cheng, M.G.; Wang, X.; Hu, Y. The emerging role of non-coding RNAs from extracellular vesicles in Alzheimer's disease. *J. Integr. Neurosci.* **2021**, *20*, 239–245. [CrossRef] [PubMed]
36. Zhang, Y.; Liu, Y.; Liu, H.; Tang, W.H. Exosomes: Biogenesis, biologic function and clinical potential. *Cell Biosci.* **2019**, *9*, 1–18. [CrossRef]
37. Wang, D.; Wang, P.; Bian, X.; Xu, S.; Zhou, Q.; Zhang, Y.; Ding, M.; Han, M.; Huang, L.; Bi, J.; et al. Elevated plasma levels of exosomal BACE1-AS combined with the volume and thickness of the right entorhinal cortex may serve as a biomarker for the detection of Alzheimer's disease. *Mol. Med. Rep.* **2020**, *22*, 227–238. [CrossRef]
38. Li, F.; Wang, Y.; Yang, H.; Xu, Y.; Zhou, X.; Zhang, X.; Xie, Z.; Bi, J. The effect of BACE1-AS on β-amyloid generation by regulating BACE1 mRNA expression. *BMC Mol. Biol.* **2019**, *20*, 1–10. [CrossRef]
39. Tan, P.-N. Receiver Operating Characteristic. In *Encyclopedia of Database Systems*; LIU, L., ÖZSU, M.T., Eds.; Springer US: Boston, MA, USA, 2009; pp. 2349–2352. ISBN 978-0-387-39940-9.
40. Gui, Y.X.; Liu, H.; Zhang, L.S.; Lv, W.; Hu, X.Y. Altered microRNA profiles in cerebrospinal fluid exosome in Parkinson disease and Alzheimer disease. *Oncotarget* **2015**, *6*, 37043–37053. [CrossRef]
41. Soreq, L.; Guffanti, A.; Salomonis, N.; Simchovitz, A.; Israel, Z.; Bergman, H.; Soreq, H. Long Non-Coding RNA and Alternative Splicing Modulations in Parkinson's Leukocytes Identified by RNA Sequencing. *PLoS Comput. Biol.* **2014**, *10*. [CrossRef]
42. Lemos, A.E.G.; Da Rocha Matos, A.; Ferreira, L.B.; Gimba, E.R.P. The long non-coding RNA PCA3: An update of its functions and clinical applications as a biomarker in prostate cancer. *Oncotarget* **2019**, *10*, 6589–6603. [CrossRef]
43. Blennow, K.; Hampel, H.; Weiner, M.; Zetterberg, H. Cerebrospinal fluid and plasma biomarkers in Alzheimer disease. *Nat. Rev. Neurol.* **2010**, *6*, 131–144. [CrossRef]
44. Bernard, D.; Prasanth, K.V.; Tripathi, V.; Colasse, S.; Nakamura, T.; Xuan, Z.; Zhang, M.Q.; Sedel, F.; Jourdren, L.; Coulpier, F.; et al. A long nuclear-retained non-coding RNA regulates synaptogenesis by modulating gene expression. *EMBO J.* **2010**, *29*, 3082–3093. [CrossRef]
45. Yao, J.; Wang, X.; Li, Y.; Shan, K.; Yang, H.; Wang, Y.; Yao, M.; Liu, C.; Li, X.; Shen, Y.; et al. Long non-coding RNA MALAT 1 regulates retinal neurodegeneration through CREB signaling. *EMBO Mol. Med.* **2016**, *8*, 1113. [CrossRef] [PubMed]
46. Zhuang, J.; Cai, P.; Chen, Z.; Yang, Q.; Chen, X.; Wang, X.; Zhuang, X. Long noncoding RNA MALAT1 and its target microRNA-125b are potential biomarkers for Alzheimer's disease management via interactions with FOXQ1, PTGS2 and CDK5. *Am. J. Transl. Res.* **2020**, *12*, 5940–5954.
47. Airavaara, M.; Pletnikova, O.; Doyle, M.E.; Zhang, Y.E.; Troncoso, J.C.; Liu, Q.R. Identification of novel GDNF isoforms and cis-antisense GDNFOS gene and their regulation in human middle temporal gyrus of Alzheimer disease. *J. Biol. Chem.* **2011**, *286*, 45093–45102. [CrossRef]
48. Straten, G.; Eschweiler, G.W.; Maetzler, W.; Laske, C.; Leyhe, T. Glial cell-line derived neurotrophic factor (GDNF) concentrations in cerebrospinal fluid and serum of patients with early Alzheimer's disease and normal controls. *J. Alzheimer's Dis.* **2009**, *18*, 331–337. [CrossRef]
49. Yuan, J.H.; Yang, F.; Wang, F.; Ma, J.Z.; Guo, Y.J.; Tao, Q.F.; Liu, F.; Pan, W.; Wang, T.T.; Zhou, C.C.; et al. A Long Noncoding RNA Activated by TGF-β promotes the invasion-metastasis cascade in hepatocellular carcinoma. *Cancer Cell* **2014**, *25*, 666–681. [CrossRef] [PubMed]
50. Ma, C.C.; Xiong, Z.; Zhu, G.N.; Wang, C.; Zong, G.; Wang, H.L.; Bian, E.B.; Zhao, B. Long non-coding RNA ATB promotes glioma malignancy by negatively regulating miR-200a. *J. Exp. Clin. Cancer Res.* **2016**, *35*, 1–13. [CrossRef] [PubMed]
51. Wang, J.; Zhou, T.; Wang, T.; Wang, B. Suppression of lncRNA-ATB prevents amyloid-β-induced neurotoxicity in PC12 cells via regulating miR-200/ZNF217 axis. *Biomed. Pharmacother.* **2018**, *108*, 707–715. [CrossRef] [PubMed]
52. Lehrer, S. Glioma and Alzheimer's Disease. *J. Alzheimer's Dis. Rep.* **2018**, *2*, 213–218. [CrossRef]

Article

Analysis of Genetic Variants Associated with Levels of Immune Modulating Proteins for Impact on Alzheimer's Disease Risk Reveal a Potential Role for SIGLEC14

Benjamin C. Shaw [1,2], Yuriko Katsumata [3], James F. Simpson [1,2], David W. Fardo [2,3] and Steven Estus [1,2,*]

1. Department of Physiology, University of Kentucky, Lexington, KY 40506, USA; benjamin.shaw@uky.edu (B.C.S.); jfsimp01@uky.edu (J.F.S.)
2. Sanders-Brown Center on Aging, University of Kentucky, Lexington, KY 40506, USA; david.fardo@uky.edu
3. Department of Biostatistics, University of Kentucky, Lexington, KY 40506, USA; katsumata.yuriko@uky.edu
* Correspondence: steve.estus@uky.edu; Tel.: +1-859-218-2388

Abstract: Genome-wide association studies (GWAS) have identified immune-related genes as risk factors for Alzheimer's disease (AD), including *TREM2* and *CD33*, frequently passing a stringent false-discovery rate. These genes either encode or signal through immunomodulatory tyrosine-phosphorylated inhibitory motifs (ITIMs) or activation motifs (ITAMs) and govern processes critical to AD pathology, such as inflammation and amyloid phagocytosis. To investigate whether additional ITIM and ITAM-containing family members may contribute to AD risk and be overlooked due to the stringent multiple testing in GWAS, we combined protein quantitative trait loci (pQTL) data from a recent plasma proteomics study with AD associations in a recent GWAS. We found that pQTLs for genes encoding ITIM/ITAM family members were more frequently associated with AD than those for non-ITIM/ITAM genes. Further testing of one family member, *SIGLEC14* which encodes an ITAM, uncovered substantial copy number variations, identified an SNP as a proxy for gene deletion, and found that gene expression correlates significantly with gene deletion. We also found that *SIGLEC14* deletion increases the expression of *SIGLEC5*, an ITIM. We conclude that many genes in this ITIM/ITAM family likely impact AD risk, and that complex genetics including copy number variation, opposing function of encoded proteins, and coupled gene expression may mask these AD risk associations at the genome-wide level.

Keywords: ITIM; ITAM; *SIGLEC14*; *SIGLEC5*; copy number variation; CNV; GWAS

1. Introduction

Genome-wide association studies (GWAS) have identified a set of polymorphisms that modulate the risk of Alzheimer's disease (AD) [1–6]. The pathways implicated in this process include innate immunity, cholesterol homeostasis, and protein trafficking [7–9]. Four of these genes, *TREM2*, *CD33*, *PILRA*, and *FCER1G*, are members of the family of non-catalytic tyrosine-phosphorylated receptors (NTRs), which function through immunomodulatory tyrosine-phosphorylated activating motifs (ITAMs) or inhibitory motifs (ITIMs). The underlying immunomodulatory pathway is further implicated by AD-associated variants in phospholipase C (*PLCG2*) and *INPP5D* which encode proteins acting downstream of these ITAM- and ITIM-containing proteins. Functional studies have informed the current hypothesis that the variants associated with AD in the ITAM/ITIM family modulate inflammation and phagocytosis [10–18].

The ITAM family, including *TREM2*, recruit kinases such as spleen tyrosine kinase (Syk) and phosphoinositide 3-kinase (PI3K) to induce downstream signaling, while the ITIM family, including *CD33*, recruit phosphatases such as SHP-1 to dephosphorylate Syk and ITAMs, thereby counteracting ITAM activity [19]. These ITAM and ITIM proteins are predominantly expressed in immune cells such as microglia. Overall, these and other studies have shown that microglia contribute to AD pathogenesis, a concept that has been reviewed recently [20–22].

The critical barrier to progress in translating GWAS candidate genes to treatments is elucidating the actions of the functional variant at the molecular level, i.e., splicing (sQTL), gene expression (eQTL), or protein level (pQTL), to understand whether the pathway affected is detrimental or beneficial to disease risk. GWAS single nucleotide polymorphisms (SNPs) in AD are frequently identified as eQTLs in the brain [23]. Sun et al. have used GWAS to identify pQTLs for the plasma proteome, including ITIM and ITAM-containing proteins [24]. To investigate the hypothesis that these pQTLs may uncover additional AD-related genes that may have been overlooked in AD GWAS because of their stringent false-discovery rate controls, we examined the Sun et al. cis-pQTL data together with the Jansen et al. AD GWAS results. Parsing the proteins from the genome-wide significant cis-pQTL dataset by whether or not an ITIM/ITAM domain was present, and then examining whether the associated SNP is nominally significant ($p < 0.05$) for AD association, found a significant overrepresentation of ITIM/ITAM encoding genes with nominal AD associations. Since one of these genes, *SIGLEC14*, has been reported to be deleted in some individuals, we investigated further and found that the pQTL and AD SNP, rs1106476, is a proxy for the previously identified deletion polymorphism [25]. We defined this deletion further by identifying additional *SIGLEC14* copy number variants and by determining the effect of *SIGLEC14* copy number on the expression of *SIGLEC14* and the neighboring *SIGLEC5*. We conclude that variants in ITIM/ITAM family members, including *SIGLEC14*, represent underappreciated potential genetic risk factors for AD.

2. Materials and Methods

2.1. Preparation of gDNA, RNA, and cDNA from Human Tissue

Human blood and anterior cingulate autopsy tissue from 61 donors were generously provided by the Sanders-Brown Alzheimer's disease center neuropathology core and have been described elsewhere [26]. The matched brain and blood samples were from deceased individuals with an average age at death of 82.4 ± 8.7 (mean $\pm$ SD) years for non-AD and 81.7 ± 6.2 years for AD subjects. The average postmortem interval (PMI) for non-AD and AD subjects was 2.8 ± 0.8 and 3.4 ± 0.6 h, respectively. Non-AD and AD samples were comprised of 48% and 55% female subjects. MMSE scores were, on average, 28.4 ± 1.6 for non-AD subjects and 11.9 ± 8.0 for AD subjects. These samples were used for genotyping and gene expression studies. Three additional blood samples matched to whole-genome sequencing (WGS) data were obtained to confirm WGS observations of additional *SIGLEC14* copies. DNA from these patients was prepared using a QIAamp DNA Blood Mini kit (Qiagen, Germantown, MD, USA) per the manufacturer's instructions.

2.2. Genotyping and Copy Number Variant Assays

Copy number variation in *SIGLEC14* was determined using a TaqMan-based copy number variant (CNV) assay (Invitrogen, Waltham, MA, USA; Catalog number 4400291, Assay number Hs03319513_cn) compared to *RNAse P* (Invitrogen, 4403326). Amplification and quantitation were performed per manufacturer instructions. Genotyping the rs1106476 was performed with a custom TaqMan assay (Invitrogen). This assay discriminates rs1106476 and rs872629, which are in perfect LD. As coinherited SNPs, this variant is also known as rs35495434.

2.3. Gene Expression by qPCR

Gene expression was quantified by qPCR with PerfeCTa SYBR Green master mix as previously described [14]. *SIGLEC14* was quantified with primers corresponding to a sequence in exons 3 and 5: 5'-CAGGTGAAACGCCAAGGAG-3' and 5'-GCGAGGAACAGGGA CTGG-3'. *SIGLEC5* was quantified with primers corresponding to sequences in exons 4 and 5: 5'-ACCATCTTCAGGAACGGCAT-3' and 5'-GGGAGCATCACAGAGCAGC-3'. Cycling conditions for all qPCRs were as follows: 95 °C, 2 min; 95 °C, 15 s, 60 °C, 15 s, 72 °C, 30 s,

40 cycles. Copy numbers present in the cDNA were determined relative to standard curves that were executed in parallel [19].

2.4. WGS Data Analysis

To investigate the frequency and range of *SIGLEC14* CNV, we performed a read-depth analysis for WGS data. We obtained compressed sequence alignment map (CRAM) files from the AD sequencing project (ADSP) and AD Neuroimaging (ADNI). We extracted paired-end reads mapped to the *SIGLEC14-SIGLEC5* locus under Genome Reference Consortium Human Build 38 (GRCh38/hg38), and then computed the depth at each position using the samtools depth function [27].

2.5. Statistical Analyses

The association of cis-pQTL proteins containing ITIM/ITAM domains and AD-associated SNPs was calculated using a simple chi-square test. Gene expression was analyzed by using JMP14 Pro using one-way analysis of variance (ANOVA) followed by Tukey's post-hoc multiple testing correction and graphed in GraphPad Prism 8.

3. Results

3.1. ITIM/ITAM pQTLs Are Overrepresented in AD GWAS Results

To evaluate whether pQTLs for ITIM or ITAM-containing proteins were associated with AD, we compiled a list of ITIM and ITAM-containing proteins from prior reviews [28–31]. The resulting list contained 187 genes and is provided as Supplemental Table S1. The cis-acting pQTLs from Sun et al. and AD associations from Jansen et al. were then matched by chromosomal coordinates [2,24]. Both datasets were provided under Genome Reference Consortium Human Build 37 (GRCh37/hg19). Genes were then subset as either coding for an ITIM/ITAM gene or not and nominally significant ($p < 0.05$) for AD association or not. The SNPs which are associated with both ITIM/ITAM protein levels in plasma and AD risk are shown in Table 1. We found that pQTLs that affect ITIM or ITAM genes were significantly overrepresented in nominally significant AD associations ($p = 6.51 \times 10^{-5}$, $\chi_1^2 = 15.95$, Table 2).

Table 1. Genes that are nominally significant for AD association with strong pQTL signal.

Gene	SNP	P (pQTL)	β (pQTL)	P (AD)	β (AD)	N (AD)	ITIM/ITAM
CD33	rs12459419	0 †	−0.94	7.13×10^{-9}	−0.01330	458,744	ITIM
FCGR3B	rs10919543	3.20×10^{-67}	0.44	0.000317	0.00806	445,293	ITAM
LILRA5	rs759819	2.50×10^{-111}	−0.54	0.00186	0.00717	454,216	ITAM
LILRB2	rs373032	7.60×10^{-146}	−0.72	0.00227	0.00763	463,880	ITIM
SIGLEC9	rs2075803	0 †	−1.23	0.00703	0.00576	466,252	ITIM
SIRPB1	rs3848788	1.20×10^{-213}	0.75	0.00942	0.00582	458,092	ITAM
COLEC12	rs2846667	9.30×10^{-12}	0.20	0.0177	0.00586	449,987	ITAM
FCRL1	rs4971155	6.30×10^{-26}	−0.26	0.0197	−0.00520	403,829	ITAM
NCR1	rs2278428	1.10×10^{-15}	−0.36	0.0249	0.00815	466,252	ITAM
SIGLEC14	rs1106476	0 †	−1.19	0.0284	0.00736	458,063	ITAM
FCRL3	rs7528684	1.40×10^{-112}	0.53	0.04	−0.00434	458,744	Both
MRC2	rs146385050	1.30×10^{-11}	−0.22	0.041	−0.00612	396,686	ITAM
SLAMF6	rs11291564	2.60×10^{-12}	0.20	0.042	−0.02450	17,477	ITAM

† The *p*-value in the analyzed summary statistics was reported as exactly 0. This does not impact our analysis, as our threshold was any cis-pQTL at $p < 0.05$.

Table 2. Overlap of pQTL and AD signals.

pQTLs	ITIM/ITAM (%)	Not ITIM/ITAM (%)	Total
AD $p < 0.05$	13 (28)	54 (10)	67
AD $p > 0.05$	34 (72)	488 (90)	522
Total	47 (100)	542 (100)	589

3.2. SIGLEC14 pQTL Is a Proxy for the Deletion Polymorphism

Previous reports have identified a *SIGLEC14* deletion [25]. Given the strong pQTL signal from rs1106476 on SIGLEC14 reported by Sun et al., and the fact that rs1106476 is within the neighboring *SIGLEC5* gene, yet has a cis-pQTL effect on SIGLEC14, we hypothesized that rs1106476 is a proxy for the *SIGLEC14* deletion polymorphism. To test this hypothesis, we genotyped a set of DNA samples for rs1106476 and quantified genomic copy number variation (CNV). We found that the proxy SNP correlates with *SIGLEC14* deletion well but not perfectly ($p < 0.0001$, $\chi_2^2 = 38.40$) (Table 3). To better understand this deletion, we then sequenced the region containing the *SIGLEC14-SIGLEC5* fusion in five minor allele carriers (two homozygous for *SIGLEC14* deletion and three heterozygous) [25]. Based on these sequencing data, relative to reference sequences, we found a 692 bp region of complete identity between *SIGLEC14* and *SIGLEC5*. Within this region, the deletion polymorphism sequence corresponds to *SIGLEC14* at the 5' end, but *SIGLEC5* on the 3' end, with respect to reference sequence data (Figure 1). Overall, this represents a 17 kb deletion.

Table 3. Evaluation of rs1106476 as a proxy for *SIGLEC14* deletion.

SIGLEC14 Copies	rs1106476 T/T	rs1106476 A/T	rs1106476 A/A	Total
0	0	1	1	2
1	6	13	0	19
2	39	0	0	39
3	2	2	0	4
Total	47	16	1	64

Blue = predicted correlation of *SIGLEC14* deletion vs. rs1106476. Each cell represents the number of DNA samples with the indicated *SIGLEC14* copy number and rs1106476 genotype.

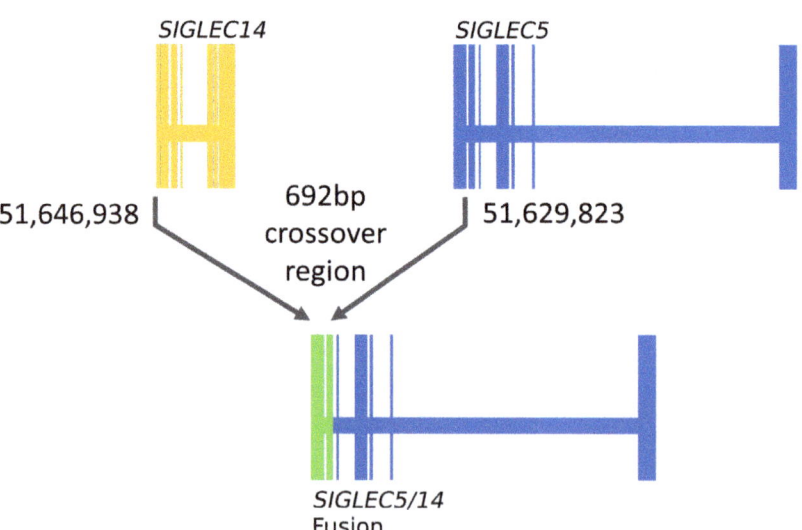

Figure 1. Identification of the *SIGLEC14* deletion site. Coordinates in both are for reference genome. Exons 1-3 of *SIGLEC14* and *SIGLEC5* are identical which confounds exact determination of the crossover event. The yellow region depicts *SIGLEC14*, the blue region depicts *SIGLEC5*, while the green region depicts the 692 bp region of complete identity where the crossover deletion occurs.

3.3. SIGLEC14 CNV Is Not Fully Captured by rs1106476

As noted in Table 3, we found some individuals that had three copies of *SIGLEC14* as detected by the CNV assay. To validate these findings, we leveraged the ADNI and ADSP WGS datasets and compared read depth in the *SIGLEC14* locus with surrounding sequences (Figure 2). Both datasets contained individuals with *SIGLEC14* copy numbers ranging from 0–3. The presence of three copies of *SIGLEC14* was cross-validated between WGS data and CNV assay in three individuals. Further, the frequencies across populations are equivalent (Table 4; $p = 6.76 \times 10^{-12}$, $\chi^2 = 69.30$). Read depths for Caucasian, African American, and other populations are shown as Supplemental Figures S1–S3.

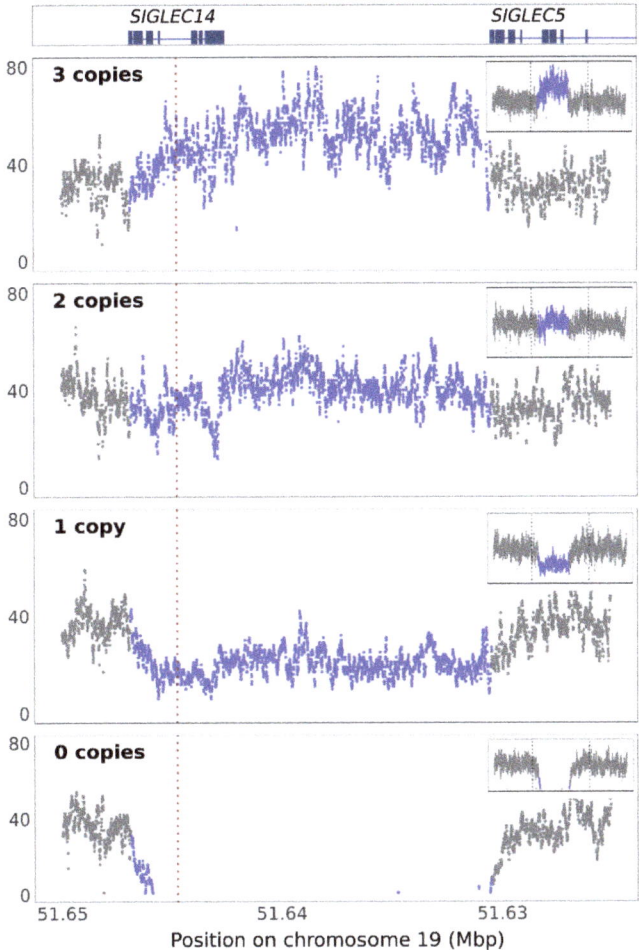

Figure 2. *SIGLEC14* CNVs detected in ADNI and ADSP cohorts. Read depth shown by chromosomal position of whole-genome sequencing in a representative example of each CNV detected. Exon/intron maps for *SIGLEC14* and *SIGLEC5* at figure top for reference. Purple: copy number variation. Inset: expanded view of locus. Red dotted line: location of copy number variation assay. The dotted line in the insets shows the boundaries of the full-size image.

Table 4. Summary of the *SIGLEC14* CNV in the 3095 sample ADSP WGS dataset.

SIGLEC14 Copy Number	Caucasian	African American	Other	Total
0	24	74	44	142
1	304	348	316	968
2	692	522	652	1866
3	21	53	43	117
4	0	1	1	2
Total	1041	998	1056	3095
Deletion MAF	0.1691	0.2485	0.1913	0.2023
Addition MAF	0.0101	0.0276	0.0213	0.0195

MAF: Minor allele frequency.

3.4. SIGLEC14 Is Expressed in Human Brain, and CNV Correlates with Gene Expression

To test whether gene expression compensation may neutralize the effect of genomic *SIGLEC14* deletion, we quantified *SIGLEC14* expression relative to *SIGLEC14* gene copy number in cDNA prepared from human brain samples. Consistent with RNAseq studies that show *SIGLEC14* is expressed in microglia, *SIGLEC14* expression strongly correlated with expression of the microglial gene *AIF1* ($p < 0.0001$, $r^2 = 0.409$, Figure 3A) [19,32]. When *SIGLEC14* expression is normalized to *AIF1* expression, *SIGLEC14* expression was dependent in a step-wise manner with *SIGLEC14* CNV ($p = 0.0002$, $F_{2,47} = 10.679$, Figure 3B). Strikingly, individuals with one copy of *SIGLEC14* have a mean *SIGLEC14* expression of 54.6% compared to individuals with two copies. We interpret this to mean that there is no compensatory increase in *SIGLEC14* expression in individuals heterozygous for *SIGLEC14* deletion.

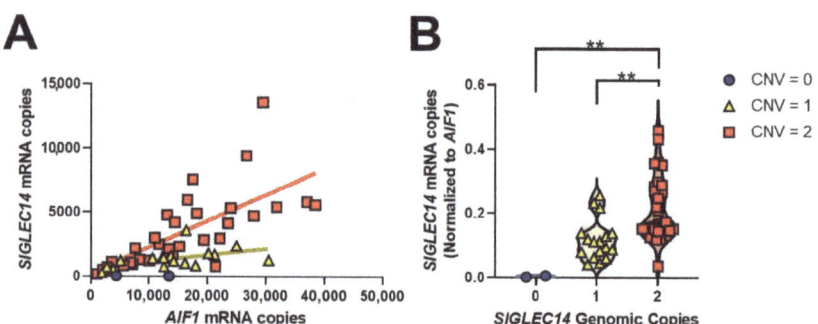

Figure 3. SIGLEC14 expression correlates with microglial gene AIF1 and SIGLEC14 CNV. (**A**) SIGLEC14 is expressed in microglia ($p < 0.0001$, $F_{1,48} = 33.19$, $r^2 = 0.409$). (**B**) SIGLEC14 CNV strongly correlates with SIGLEC14 gene expression ($p = 0.0002$, $F_{2,47} = 10.679$), Tukey's post-hoc multiple comparisons test. ** $p < 0.01$. We do not have statistical power to compare expression with CNV > 2, given its low MAF.

3.5. SIGLEC14 Deletion Leads to Increased SIGLEC5 Expression

To test whether *SIGLEC5* expression changed with respect to *SIGLEC14* deletion, we quantified *SIGLEC5* expression relative to *SIGLEC14* CNV in these same brain samples. Since *SIGLEC5* does not have its own promoter and there are no H3K27 acetylation peaks between *SIGLEC14* and *SIGLEC5*, we hypothesized that an inverse relationship exists between *SIGLEC14* CNV and *SIGLEC5* expression, where a *SIGLEC14* deletion brings *SIGLEC5* closer to the promoter leading to increased transcription (Supplemental Figure S4) [33–35]. We found that *SIGLEC5* expression significantly increases with respect to *SIGLEC14* genomic deletions (Figure 4; $p = 0.0220$, $F_{2,46} = 4.151$).

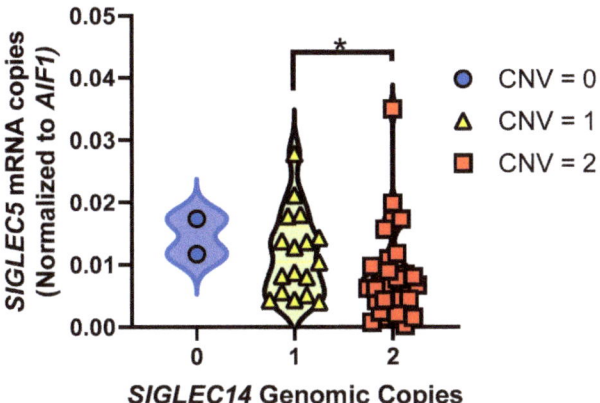

Figure 4. *SIGLEC5* expression inversely correlates with *SIGLEC14* CNV. *SIGLEC5* expression increases with fewer copies of *SIGLEC14*, presumably due to proximity to regulatory elements ($p = 0.0220$, $F_{2,46} = 4.151$), Tukey's post-hoc multiple comparisons test. * $p = 0.0389$.

4. Discussion

The primary finding of this paper is that pQTLs for ITIM and ITAM-containing proteins are overrepresented as being nominally significant for AD risk, suggesting that the ITIM and ITAM family of proteins may contribute to AD pathogenesis. This adds to the current body of work which supports the hypothesis that AD is mediated, at least in part, by immune cell dysfunction [1,4,5,36]. Indeed, transcriptomics and genomics studies have frequently identified genes predominantly expressed in microglia within the CNS as associated with AD risk [37–41]. Within a pQTL study, variants that affect the expression of the ITIM/ITAM family of genes—which govern immune cell activation state—are more commonly associated with AD risk than variants for genes, not in this family (Table 2). Although we hypothesized that variants that enhanced ITAM levels or decreased ITIM levels would be associated with reduced AD risk, this was not observed. This likely indicates that while some of these pQTLs may reflect increased functional signaling, others may involve alterations in splicing to generate soluble isoforms or may increase susceptibility to cleavage from the cell surface. Hence, an SNP that associates with increased plasma protein levels does not necessarily correlate with increased cell surface expression and signaling.

SIGLEC14 was selected for further investigation based on its previously reported deletion polymorphism and close relationship to another AD-associated gene, *CD33* [2,25]. Since SNPs have previously been recognized as proxies for deletion of other genes [42–44], and SIGLEC14 deletion has been previously reported [25], we hypothesized that the strong pQTL signal from rs1106476 reported in Sun et al. [24] correlated with *SIGLEC14* deletion. Indeed, we found that rs1106476 is a proxy for *SIGLEC14* deletion and the minor allele count corresponds to the number of *SIGLEC14* deletions in 89% of cases in our dataset (Table 3).

This proxy variant does not, however, predict copy numbers greater than two. For instance, we observed four individuals with three copies of *SIGLEC14*; two of these individuals were homozygous minor for rs1106476 and two were heterozygous for rs1106476 (Table 3). Additional copy number variation is also present in the ADSP and ADNI sequencing projects (Figure 2). These CNVs are equivalent across populations in these datasets (Table 4, Supplemental Figures S1–S3). Based on these data and the recombination peak which spans from upstream of *SIGLEC14* through exon 8 of *SIGLEC5* (Supplemental Figure S5), we hypothesize that the additional copies integrate from a deletion event, though far less frequently than the deletion itself [45]. Across the 3095 individual WGS dataset in ADSP, we found *SIGLEC14* deletion has a minor allele frequency (MAF) of 0.2023, while insertion occurs at a MAF of only 0.0195, suggesting a 10-times lower rate of integration than deletion (Table 4).

In the brain, *SIGLEC14* is predominantly expressed in microglia, in keeping with its putative role as an immune receptor (Figure 3A). The *SIGLEC14* deletion polymorphism also strongly correlates with *SIGLEC14* gene expression (Figure 3B). Due to the low frequency of the additional copy integration, we do not have sufficient samples with which to correlate *SIGLEC14* expression to additional copy numbers, nor can we conclude whether additional *SIGLEC14* genomic copies are transcribed in frame and subsequently produce protein.

We also find that *SIGLEC14* deletion increases the expression of *SIGLEC5* (Figure 4). For individuals with at least one copy of *SIGLEC14*, the expression of *SIGLEC14* is substantially higher than *SIGLEC5*. Coupled with the lack of an independent promoter or H3K27 acetylation peaks between the two genes in GeneHancer or Encode, respectively, we infer that expression of both genes is governed by a common promoter proximal to *SIGLEC14*, that the integrity of this promoter is preserved after *SIGLEC14* deletion, and that *SIGLEC14* deletion results in an increase in *SIGLEC5* expression due to its closer proximity to this common element. The SIGLEC family of receptors bind sialic acids as ligands to initiate their signaling cascades, and sialylated proteins, as well as gangliosides, are abundant in amyloid plaques [46–48]. This decrease in expression of *SIGLEC14*, an ITAM-coupling protein, and concomitant increase in expression of *SIGLEC5*, an ITIM-containing protein, may lead to a dampened microglial activation state or proportion of activated microglia in deletion carriers. We speculate that decreased SIGLEC14 expression and increased SIGLEC5 expression may decrease the phagocytic capacity in AD. This is similar to the inverse relationship between *TREM2* and *CD33*, two well-known AD risk factors. Loss of the ITAM-containing TREM2 decreases phagocytic capacity, while loss of CD33 increases phagocytic capacity [11,13,49]. Since *TREM2*, which couples with DAP12, is critical for the transition of microglia into a full disease-associated phenotype, *SIGLEC14* may also contribute to this transition [50]. Future studies could investigate whether at the single-cell level *SIGLEC14* CNV affects disease-associated microglial induction.

Copy number variation may represent a relatively unexplored source of genetic variation in AD [51]. GWAS such as Jansen et al. rely on SNPs, which do not always capture the full range of variation [2]. Additionally, "camouflaged" genes such as *SIGLEC5* and *SIGLEC14* with high sequence identity due to gene duplication are challenging for WGS and WES technologies which rely on small fragments of DNA sequence, typically under 250 bp reads [51]. As such, variants which may have disease relevance and association may be overlooked with current methods. *SIGLEC14* is an example of one such possibly overlooked risk contributor in AD. *SIGLEC14* encodes an ITAM protein and signals through DAP12 similar to *TREM2*, and deletion of *SIGLEC14* is associated with increased AD risk, also similar to SNPs that reduce *TREM2* function [1,3–5]. Ligands for SIGLEC14, which include sialylated proteins, are commonly found within amyloid plaques similar to ligands for TREM2. We propose that the effect size and significance of association are masked through copy number variation not accounted for using the proxy SNP alone, i.e., loss of SIGLEC14 function likely increases risk, but the proxy SNP rs1106476 occasionally also marks the individuals with an extra *SIGLEC14* copy, thus reducing the power of rs1106476 association with AD. We thus conclude that *SIGLEC14* represents a potentially overlooked AD genetic risk factor due to complex genetics.

Supplementary Materials: The following are available online at https://www.mdpi.com/article/10.3390/genes12071008/s1, Figure S1: Whole genome sequencing (WGS) read depth data from the Alzheimer's Disease Sequencing Project (ASDP) in Caucasian population, Figure S2: WGS read depth data from the ASDP in African American population, Figure S3: WGS read depth data from the ASDP in all other populations, Figure S4: The *SIGLEC14* locus contains no H3K27Ac peaks nor regulatory elements between *SIGLEC14* and *SIGLEC5*. Expression of *SIGLEC14* is approximately ten times higher than *SIGLEC5* in individuals with both copies of *SIGLEC14*, while *SIGLEC5* expression is higher in individuals lacking *SIGLEC14* copies, in keeping with a common promoter or enhancer governing the single locus, Figure S5: *SIGLEC5* and *SIGLEC14* share a broad recombination peak (gray line). Note that, since *SIGLEC14* and *SIGLEC5* are on the minus strand, these genes appear inverted in this figure and read right-to-left, Table S1: List of ITIM/ITAM genes and their aliases.

Author Contributions: Conceptualization, B.C.S., D.W.F. and S.E.; data curation, B.C.S. and Y.K.; formal analysis, B.C.S., Y.K., D.W.F. and S.E.; funding acquisition, D.W.F. and S.E.; investigation, B.C.S., Y.K., J.F.S., D.W.F. and S.E.; methodology, B.C.S., Y.K., J.F.S., D.W.F. and S.E.; resources, B.C.S., Y.K., J.F.S., D.W.F. and S.E.; supervision, D.W.F. and S.E.; visualization, B.C.S. and Y.K.; writing—original draft, B.C.S.; writing—reviewing and editing, B.C.S., Y.K., J.F.S., D.W.F. and S.E. All authors have read and agreed to the published version of the manuscript.

Funding: This work was funded by grants R21AG068370 (S.E.), RF1AG059717 (S.E.), RF1AG059717-01S1 (S.E. & B.C.S.), R56AG057191 (D.W.F. & Y.K.), R01AG057187 (D.W.F. & Y.K.), R21AG061551 (D.W.F. & Y.K.), R01AG054060 (D.W.F. & Y.K.), and the UK-ADC P30AG028383 from the National Institute on Aging.

Institutional Review Board Statement: The study was conducted according to the guidelines of the Declaration of Helsinki and approved by the Institutional Review Board at the University of Kentucky (protocol code 48095 on 9/16/2020).

Informed Consent Statement: Informed consent was obtained for all subjects involved in the study.

Data Availability Statement: The Sun et al. proteomics dataset is available through the supplementary materials provided in the original publication, accessed on 30 January 2020 [24]. The Jansen et al. AD summary statistics are available through: https://ctg.cncr.nl/software/summary_statistics, accessed on 10 January 2019.

Conflicts of Interest: The authors declare no conflict of interest.

References

1. Hollingworth, P.; Harold, D.; Sims, R.; Gerrish, A.; Lambert, J.C.; Carrasquillo, M.M.; Abraham, R.; Hamshere, M.L.; Pahwa, J.S.; Moskvina, V.; et al. Common variants at ABCA7, MS4A6A/MS4A4E, EPHA1, CD33 and CD2AP are associated with Alzheimer's disease. *Nat. Genet.* **2011**, *43*, 429–435. [CrossRef] [PubMed]
2. Jansen, I.E.; Savage, J.E.; Watanabe, K.; Bryois, J.; Williams, D.M.; Steinberg, S.; Sealock, J.; Karlsson, I.K.; Hagg, S.; Athanasiu, L.; et al. Genome-wide meta-analysis identifies new loci and functional pathways influencing Alzheimer's disease risk. *Nat. Genet.* **2019**, *51*, 404–413. [CrossRef] [PubMed]
3. Kunkle, B.W.; Grenier-Boley, B.; Sims, R.; Bis, J.C.; Damotte, V.; Naj, A.C.; Boland, A.; Vronskaya, M.; van der Lee, S.J.; Amlie-Wolf, A.; et al. Genetic meta-analysis of diagnosed Alzheimer's disease identifies new risk loci and implicates Abeta, tau, immunity and lipid processing. *Nat. Genet.* **2019**, *51*, 414–430. [CrossRef] [PubMed]
4. Lambert, J.C.; Ibrahim-Verbaas, C.A.; Harold, D.; Naj, A.C.; Sims, R.; Bellenguez, C.; DeStafano, A.L.; Bis, J.C.; Beecham, G.W.; Grenier-Boley, B.; et al. Meta-analysis of 74,046 individuals identifies 11 new susceptibility loci for Alzheimer's disease. *Nat. Genet.* **2013**, *45*, 1452–1458. [CrossRef]
5. Naj, A.C.; Jun, G.; Beecham, G.W.; Wang, L.S.; Vardarajan, B.N.; Buros, J.; Gallins, P.J.; Buxbaum, J.D.; Jarvik, G.P.; Crane, P.K.; et al. Common variants at MS4A4/MS4A6E, CD2AP, CD33 and EPHA1 are associated with late-onset Alzheimer's disease. *Nat. Genet.* **2011**, *43*, 436–441. [CrossRef]
6. Novikova, G.; Kapoor, M.; Tcw, J.; Abud, E.M.; Efthymiou, A.G.; Chen, S.X.; Cheng, H.; Fullard, J.F.; Bendl, J.; Liu, Y.; et al. Integration of Alzheimer's disease genetics and myeloid genomics identifies disease risk regulatory elements and genes. *Nat. Commun.* **2021**, *12*, 1–14. [CrossRef]
7. Jones, L.; Holmans, P.A.; Hamshere, M.L.; Harold, D.; Moskvina, V.; Ivanov, D.; Pocklington, A.; Abraham, R.; Hollingworth, P.; Sims, R.; et al. Genetic evidence implicates the immune system and cholesterol metabolism in the aetiology of Alzheimer's disease. *PLoS ONE* **2010**, *5*, e13950. [CrossRef]
8. Karch, C.M.; Goate, A.M. Alzheimer's disease risk genes and mechanisms of disease pathogenesis. *Biol. Psychiatry* **2015**, *77*, 43–51. [CrossRef]
9. Malik, M.; Parikh, I.; Vasquez, J.B.; Smith, C.; Tai, L.; Bu, G.; LaDu, M.J.; Fardo, D.W.; Rebeck, G.W.; Estus, S. Genetics ignite focus on microglial inflammation in Alzheimer's disease. *Mol. Neurodegener.* **2015**, *10*, 52. [CrossRef] [PubMed]
10. Bhattacherjee, A.; Jung, S.J.; Ho, M.; Eskandari-Sedighi, G.; St. Laurent, C.D.; McCord, K.A.; Bains, A.; Gaurav, S.; Sarkar, S.S.; Plemel, J.; et al. The CD33 short isoform is a gain-of-function variant that enhances Aβ1-42 phagocytosis in microglia. *Mol. Neurodegener.* **2021**, *16*, 1–22. [CrossRef]
11. Bhattacherjee, A.; Rodrigues, E.; Jung, J.; Luzentales-Simpson, M.; Enterina, J.R.; Galleguillos, D.; St. Laurent, C.D.; Nakhaei-Nejad, M.; Fuchsberger, F.F.; Streith, L.; et al. Repression of phagocytosis by human CD33 is not conserved with mouse CD33. *Commun. Biol.* **2019**, *2*, 450. [CrossRef]

12. Chan, G.; White, C.C.; Winn, P.A.; Cimpean, M.; Replogle, J.M.; Glick, L.R.; Cuerdon, N.E.; Ryan, K.J.; Johnson, K.A.; Schneider, J.A.; et al. CD33 modulates TREM2: Convergence of Alzheimer loci. *Nat. Neurosci.* **2015**, *18*, 1556–1558. [CrossRef]
13. Griciuc, A.; Serrano-Pozo, A.; Parrado, A.R.; Lesinski, A.N.; Asselin, C.N.; Mullin, K.; Hooli, B.; Choi, S.H.; Hyman, B.T.; Tanzi, R.E. Alzheimer's Disease Risk Gene CD33 Inhibits Microglial Uptake of Amyloid Beta. *Neuron* **2013**, *78*, 631–643. [CrossRef] [PubMed]
14. Malik, M.; Chiles, I.I.I.J.; Xi, H.S.; Medway, C.; Simpson, J.; Potluri, S.; Howard, D.; Liang, Y.; Paumi, C.M.; Mukherjee, S.; et al. Genetics of CD33 in Alzheimer's disease and acute myeloid leukemia. *Hum. Mol. Genet.* **2015**, *24*, 3557–3570. [CrossRef]
15. Malik, M.; Simpson, J.F.; Parikh, I.; Wilfred, B.R.; Fardo, D.W.; Nelson, P.T.; Estus, S. CD33 Alzheimer's Risk-Altering Polymorphism, CD33 Expression, and Exon 2 Splicing. *J. Neurosci.* **2013**, *33*, 13320–13325. [CrossRef] [PubMed]
16. Raj, T.; Ryan, K.J.; Replogle, J.M.; Chibnik, L.B.; Rosenkrantz, L.; Tang, A.; Rothamel, K.; Stranger, B.E.; Bennett, D.A.; Evans, D.A.; et al. CD33: Increased inclusion of exon 2 implicates the Ig V-set domain in Alzheimer's disease susceptibility. *Hum. Mol. Genet.* **2014**, *23*, 2729–2736. [CrossRef] [PubMed]
17. Siddiqui, S.S.; Springer, S.A.; Verhagen, A.; Sundaramurthy, V.; Alisson-Silva, F.; Jiang, W.; Ghosh, P.; Varki, A. The Alzheimer's disease-protective CD33 splice variant mediates adaptive loss of function via diversion to an intracellular pool. *J. Biol. Chem.* **2017**, *292*, 15312–15320. [CrossRef]
18. McQuade, A.; Kang, Y.J.; Hasselmann, J.; Jairaman, A.; Sotelo, A.; Coburn, M.; Shabestari, S.K.; Chadarevian, J.P.; Fote, G.; Tu, C.H.; et al. Gene expression and functional deficits underlie TREM2-knockout microglia responses in human models of Alzheimer's disease. *Nat. Commun.* **2020**, *11*, 1–17. [CrossRef] [PubMed]
19. Estus, S.; Shaw, B.C.; Devanney, N.; Katsumata, Y.; Press, E.E.; Fardo, D.W. Evaluation of CD33 as a genetic risk factor for Alzheimer's disease. *Acta Neuropathol.* **2019**, *138*, 187–199. [CrossRef]
20. Griciuc, A.; Tanzi, R.E. The role of innate immune genes in Alzheimer's disease. *Curr. Opin. Neurol.* **2021**, *34*, 228–236. [CrossRef]
21. Efthymiou, A.G.; Goate, A.M. Late onset Alzheimer's disease genetics implicates microglial pathways in disease risk. *Mol. Neurodegener.* **2017**, *12*, 43. [CrossRef]
22. Gandy, S.; Heppner, F.L. Microglia as dynamic and essential components of the amyloid hypothesis. *Neuron* **2013**, *78*, 575–577. [CrossRef]
23. Allen, M.; Zou, F.; Chai, H.S.; Younkin, C.S.; Crook, J.; Pankratz, V.S.; Carrasquillo, M.M.; Rowley, C.N.; Nair, A.A.; Middha, S.; et al. Novel late-onset Alzheimer disease loci variants associate with brain gene expression. *Neurology* **2012**, *79*, 221–228. [CrossRef] [PubMed]
24. Sun, B.B.; Maranville, J.C.; Peters, J.E.; Stacey, D.; Staley, J.R.; Blackshaw, J.; Burgess, S.; Jiang, T.; Paige, E.; Surendran, P.; et al. Genomic atlas of the human plasma proteome. *Nature* **2018**, *558*, 73–79. [CrossRef]
25. Yamanaka, M.; Kato, Y.; Angata, T.; Narimatsu, H. Deletion polymorphism of SIGLEC14 and its functional implications. *Glycobiology* **2009**, *19*, 841–846. [CrossRef]
26. Zou, F.; Gopalraj, R.K.; Lok, J.; Zhu, H.; Ling, I.F.; Simpson, J.F.; Tucker, H.M.; Kelly, J.F.; Younkin, S.G.; Dickson, D.W.; et al. Sex-dependent association of a common low-density lipoprotein receptor polymorphism with RNA splicing efficiency in the brain and Alzheimer's disease. *Hum. Mol. Genet.* **2007**, *17*, 929–935. [CrossRef] [PubMed]
27. Li, H.; Handsaker, B.; Wysoker, A.; Fennell, T.; Ruan, J.; Homer, N.; Marth, G.; Abecasis, G.; Durbin, R.; Genome Project Data Processing Subgroup. The Sequence Alignment/Map format and SAMtools. *Bioinformatics* **2009**, *25*, 2078–2079. [CrossRef] [PubMed]
28. Barrow, A.D.; Trowsdale, J. You say ITAM and I say ITIM, let's call the whole thing off: The ambiguity of immunoreceptor signalling. *Eur. J. Immunol.* **2006**, *36*, 1646–1653. [CrossRef]
29. Dushek, O.; Goyette, J.; van der Merwe, P.A. Non-catalytic tyrosine-phosphorylated receptors. *Immunol. Rev.* **2012**, *250*, 258–276. [CrossRef]
30. Isakov, N. Immunoreceptor tyrosine-based activation motif (ITAM), a unique module linking antigen and Fc receptors to their signaling cascades. *J. Leukoc. Biol.* **1997**, *61*, 6–16. [CrossRef]
31. Ravetch, J.V. Immune Inhibitory Receptors. *Science* **2000**, *290*, 84–89. [CrossRef]
32. Zhang, Y.; Sloan, S.A.; Clarke, L.E.; Caneda, C.; Plaza, C.A.; Blumenthal, P.D.; Vogel, H.; Steinberg, G.K.; Edwards, M.S.; Li, G.; et al. Purification and Characterization of Progenitor and Mature Human Astrocytes Reveals Transcriptional and Functional Differences with Mouse. *Neuron* **2016**, *89*, 37–53. [CrossRef]
33. An integrated encyclopedia of DNA elements in the human genome. *Nature* **2012**, *489*, 57–74. [CrossRef]
34. Fishilevich, S.; Nudel, R.; Rappaport, N.; Hadar, R.; Plaschkes, I.; Iny Stein, T.; Rosen, N.; Kohn, A.; Twik, M.; Safran, M.; et al. GeneHancer: Genome-wide integration of enhancers and target genes in GeneCards. *Database* **2017**, *2017*. [CrossRef]
35. Kent, W.J.; Sugnet, C.W.; Furey, T.S.; Roskin, K.M.; Pringle, T.H.; Zahler, A.M.; Haussler, A.D. The Human Genome Browser at UCSC. *Genome Res.* **2002**, *12*, 996–1006. [CrossRef]
36. Mawuenyega, K.G.; Sigurdson, W.; Ovod, V.; Munsell, L.; Kasten, T.; Morris, J.C.; Yarasheski, K.E.; Bateman, R.J. Decreased Clearance of CNS-Amyloid in Alzheimer's Disease. *Science* **2010**, *330*, 1774. [CrossRef]
37. Holtman, I.R.; Raj, D.D.; Miller, J.A.; Schaafsma, W.; Yin, Z.; Brouwer, N.; Wes, P.D.; Möller, T.; Orre, M.; Kamphuis, W.; et al. Induction of a common microglia gene expression signature by aging and neurodegenerative conditions: A co-expression meta-analysis. *Acta Neuropathol. Commun.* **2015**, *3*, 31. [CrossRef] [PubMed]

38. Miller, J.A.; Woltjer, R.L.; Goodenbour, J.M.; Horvath, S.; Geschwind, D.H. Genes and pathways underlying regional and cell type changes in Alzheimer's disease. *Genome Med.* **2013**, *5*, 48. [CrossRef] [PubMed]
39. Orre, M.; Kamphuis, W.; Osborn, L.M.; Melief, J.; Kooijman, L.; Huitinga, I.; Klooster, J.; Bossers, K.; Hol, E.M. Acute isolation and transcriptome characterization of cortical astrocytes and microglia from young and aged mice. *Neurobiol. Aging* **2014**, *35*, 1–14. [CrossRef] [PubMed]
40. Wes, P.D.; Easton, A.; Corradi, J.; Barten, D.M.; Devidze, N.; Decarr, L.B.; Truong, A.; He, A.; Barrezueta, N.X.; Polson, C.; et al. Tau Overexpression Impacts a Neuroinflammation Gene Expression Network Perturbed in Alzheimer's Disease. *PLoS ONE* **2014**, *9*, e106050. [CrossRef]
41. Zhang, B.; Gaiteri, C.; Bodea, L.-G.; Wang, Z.; McElwee, J.; Podtelezhnikov, A.A.; Zhang, C.; Xie, T.; Tran, L.; Dobrin, R.; et al. Integrated Systems Approach Identifies Genetic Nodes and Networks in Late-Onset Alzheimer's Disease. *Cell* **2013**, *153*, 707–720. [CrossRef] [PubMed]
42. Abdollahi, M.R.; Huang, S.; Rodriguez, S.; Guthrie, P.A.I.; Smith, G.D.; Ebrahim, S.; Lawlor, D.A.; Day, I.N.M.; Gaunt, T.R. Homogeneous Assay of rs4343, anACEI/D Proxy, and an Analysis in the British Women's Heart and Health Study (BWHHS). *Dis. Markers* **2008**, *24*, 11–17. [CrossRef]
43. Hinds, D.A.; Kloek, A.P.; Jen, M.; Chen, X.; Frazer, K.A. Common deletions and SNPs are in linkage disequilibrium in the human genome. *Nat. Genet.* **2006**, *38*, 82–85. [CrossRef]
44. McCarroll, S.A.; Huett, A.; Kuballa, P.; Chilewski, S.D.; Landry, A.; Goyette, P.; Zody, M.C.; Hall, J.L.; Brant, S.R.; Cho, J.H.; et al. Deletion polymorphism upstream of IRGM associated with altered IRGM expression and Crohn's disease. *Nat. Genet.* **2008**, *40*, 1107–1112. [CrossRef] [PubMed]
45. Machiela, M.J.; Chanock, S.J. LDlink: A web-based application for exploring population-specific haplotype structure and linking correlated alleles of possible functional variants. *Bioinformatics* **2015**, *31*, 3555–3557. [CrossRef]
46. Ariga, T.; McDonald, M.P.; Yu, R.K. Thematic Review Series: Sphingolipids. Role of ganglioside metabolism in the pathogenesis of Alzheimer's disease—A review. *J. Lipid Res.* **2008**, *49*, 1157–1175. [CrossRef] [PubMed]
47. Yanagisawa, K. Role of gangliosides in Alzheimer's disease. *Biochim. Biophys. Acta (BBA) Biomembr.* **2007**, *1768*, 1943–1951. [CrossRef]
48. Salminen, A.; Kaarniranta, K. Siglec receptors and hiding plaques in Alzheimer's disease. *J. Mol. Med.* **2009**, *87*, 697–701. [CrossRef]
49. Griciuc, A.; Patel, S.; Federico, A.N.; Choi, S.H.; Innes, B.J.; Oram, M.K.; Cereghetti, G.; McGinty, D.; Anselmo, A.; Sadreyev, R.I.; et al. TREM2 Acts Downstream of CD33 in Modulating Microglial Pathology in Alzheimer's Disease. *Neuron* **2019**, *103*, 820–835.e827. [CrossRef]
50. Keren-Shaul, H.; Spinrad, A.; Weiner, A.; Matcovitch-Natan, O.; Dvir-Szternfeld, R.; Ulland, T.K.; David, E.; Baruch, K.; Lara-Astaiso, D.; Toth, B.; et al. A Unique Microglia Type Associated with Restricting Development of Alzheimer's Disease. *Cell* **2017**, *169*, 1276–1290.e1217. [CrossRef]
51. Ebbert, M.T.W.; Jensen, T.D.; Jansen-West, K.; Sens, J.P.; Reddy, J.S.; Ridge, P.G.; Kauwe, J.S.K.; Belzil, V.; Pregent, L.; Carrasquillo, M.M.; et al. Systematic analysis of dark and camouflaged genes reveals disease-relevant genes hiding in plain sight. *Genome. Biol.* **2019**, *20*, 97. [CrossRef] [PubMed]

Article

TOMM40 RNA Transcription in Alzheimer's Disease Brain and Its Implication in Mitochondrial Dysfunction

Eun-Gyung Lee [1], Sunny Chen [1], Lesley Leong [1], Jessica Tulloch [1] and Chang-En Yu [1,2,*]

[1] Geriatric Research, Education, and Clinical Center, VA Puget Sound Health Care System, Seattle, WA 98108, USA; eun-gyung.lee@va.gov (E.-G.L.); sunny.chen@va.gov (S.C.); lesley.leong@va.gov (L.L.); jessica.tulloch@va.gov (J.T.)
[2] Department of Medicine, University of Washington, Seattle, WA 98195, USA
* Correspondence: changeyu@uw.edu

Citation: Lee, E.-G.; Chen, S.; Leong, L.; Tulloch, J.; Yu, C.-E. TOMM40 RNA Transcription in Alzheimer's Disease Brain and Its Implication in Mitochondrial Dysfunction. *Genes* 2021, 12, 871. https://doi.org/10.3390/genes12060871

Academic Editors: Laura Ibanez and Justin Miller

Received: 14 May 2021
Accepted: 4 June 2021
Published: 6 June 2021

Publisher's Note: MDPI stays neutral with regard to jurisdictional claims in published maps and institutional affiliations.

Copyright: © 2021 by the authors. Licensee MDPI, Basel, Switzerland. This article is an open access article distributed under the terms and conditions of the Creative Commons Attribution (CC BY) license (https://creativecommons.org/licenses/by/4.0/).

Abstract: Increasing evidence suggests that the Translocase of Outer Mitochondria Membrane 40 (*TOMM40*) gene may contribute to the risk of Alzheimer's disease (AD). Currently, there is no consensus as to whether *TOMM40* expression is up- or down-regulated in AD brains, hindering a clear interpretation of *TOMM40*'s role in this disease. The aim of this study was to determine if *TOMM40* RNA levels differ between AD and control brains. We applied RT-qPCR to study *TOMM40* transcription in human postmortem brain (PMB) and assessed associations of these RNA levels with genetic variants in *APOE* and *TOMM40*. We also compared *TOMM40* RNA levels with mitochondrial functions in human cell lines. Initially, we found that the human genome carries multiple *TOMM40* pseudogenes capable of producing highly homologous RNAs that can obscure precise *TOMM40* RNA measurements. To circumvent this obstacle, we developed a novel RNA expression assay targeting the primary transcript of *TOMM40*. Using this assay, we showed that *TOMM40* RNA was upregulated in AD PMB. Additionally, elevated *TOMM40* RNA levels were associated with decreases in mitochondrial DNA copy number and mitochondrial membrane potential in oxidative stress-challenged cells. Overall, differential transcription of *TOMM40* RNA in the brain is associated with AD and could be an indicator of mitochondrial dysfunction.

Keywords: *TOMM40* gene; Alzheimer's disease; RNA transcription; pseudogene; mitochondrial dysfunction

1. Introduction

Understanding the role that genetics plays in the pathogenesis of AD has been a major research focus for the past three decades. These collective efforts have provided valuable insights into the molecular mechanisms associated with this disease. The advancement of genome-wide approaches has led to the identification of more than 40 AD-associated genetic loci. However, most of these loci have only moderate effect sizes with odds ratios ranging from 1.1 to 1.5 (AlzGene), except for the apolipoprotein E gene (*APOE*) which has an odds ratio of 3.7. The strength of the association between *APOE* and AD risk is orders of magnitude larger than all other AD loci combined, suggesting that this locus is a major biological contributor to the risk of AD. Therefore, deciphering the mechanistic role of the *APOE* locus in AD should provide insight into the etiology of this devastating disease.

Besides *APOE* itself, the extended region surrounding *APOE* has also been consistently identified by genome-wide association studies (GWAS) to strongly associate with AD [1–4]. This extended region consists of at least three additional genes (i.e., *NECTIN2*, *TOMM40*, and Apolipoprotein C1 (*APOC1*)), which carry out specific cellular functions that may possibly intersect with AD pathophysiology. Because of the strong linkage disequilibrium (LD) of these genes with *APOE*, researchers have always assumed that the disease-associated genetic signals from these genes solely reflect their associations with the *APOE* ε4 allele. However, increasing evidence points to a different interpretation. For example, a genome-wide linkage study of 71 Swedish late onset AD families found that the strongest signal in a multipoint linkage analysis of *APOE* ε4-negative families still resided in the *APOE* region [5]. Furthermore, multiple studies have shown that individuals who carry an African ε4 haplotype of *APOE* have less risk of developing AD when compared to those with a Caucasian ε4 haplotype [6–8]. These observations suggest the presence of loci in this region, beyond *APOE*, that may influence AD risk. One strong candidate is the *TOMM40* gene.

TOMM40 encodes a mitochondrial channel protein TOM40, which is essential for the formation of a translocase of the mitochondrial outer membrane (TOM) complex [9]. The TOM complex is involved in the recognition and import of nuclear-encoded proteins into the mitochondria [10]. Alterations of mitochondrial metabolism have gradually been accepted as prominent features in AD and mitochondrial dysfunction is a known characteristic of the disease [11–14]. Mitochondrial degeneration has shown to be an early sign of AD pathology, appearing even before neurofibrillary tangles (NFT) [15]. Damages to both the components and structure of mitochondria are extensively reported in AD [16], and the deficiency of several key antioxidant enzymes is a well-established hallmark of the AD brain [17]. Thus, abnormal mitochondrial dynamics, including components, morphology, membrane potential, and DNA copy number could contribute to AD risk [15,18].

TOMM40 has not only been genetically linked to AD risk but may also be functionally connected with AD pathophysiology. In a Chinese cohort, SNPs in *TOMM40* remained statistically significantly associated with AD after adjusting for age, sex, and *APOE* ε4 status [19]. A deoxythymidine homopolymer (poly-T) at rs10524523 within intron 6 of the *TOMM40* has been associated with the risk and age at onset of AD [20–23]; and the "VL" variant of this poly-T marker has been associated with increased mRNA expression of both *TOMM40* and *APOE* in *APOE* ε3/ε3 brain [24]. In addition, our own high-density SNP association studies identified genetic variants in *TOMM40* to be strongly associated with AD in Caucasians, after controlling for the *APOE* ε2/ε3/ε4 alleles [25,26]. Our quantitative trait loci studies showed that there is an association between *TOMM40* SNPs and apoE protein levels in both cerebrospinal fluid and PMB, suggesting that genetic variation within *TOMM40* may be associated with *APOE* and *TOMM40* expression in the human brain [27–29]. Furthermore, there is evidence supporting a direct connection between *TOMM40* and Aβ activity. For example, the Aβ peptide is imported into the mitochondria via the TOM40 protein [30] and the amyloid precursor protein has been reported to be associated with TOM40 in AD, but not controls [31]; Aβ peptides and mis-directed amyloid precursor protein interfere with mitochondrial protein import and disrupt mitochondrial function [31–33]; and the accumulation of Aβ in mitochondria leads to the overproduction of reactive oxygen species [30,34]. Given that *TOMM40* appears to be involved in APP/Aβ translocation and metabolism as well as *APOE* regulation, it is plausible that *TOMM40* plays a role in AD via effects on mitochondrial function. Consequently, *TOMM40* expression levels may be impacting mitochondrial function and contributing to AD risk.

Expression of *TOMM40* has been investigated in peripheral blood. Numerous studies consistently showed lower *TOMM40* mRNA levels in AD blood samples compared to controls [35–38], and a decrease in TOM40 protein level has also been observed in AD blood [37]. However, studies using human PMB are scarce and have generated conflicting results. For example, *TOMM40* mRNA levels were reported as both increased and decreased in the AD frontal cortex [39], or significantly increased in AD temporal and occipital cortices [24]. Currently, there is no consensus as to whether *TOMM40* gene expression is up- or down-regulated in AD brains, and this inconsistency hinders a unified clear interpretation of *TOMM40*'s role in AD risk. The aim of this study was to definitively determine if PMB *TOMM40* mRNA levels differ between AD and control subjects.

2. Materials and Methods

2.1. Human PMB and Cell Lines

This work used deidentified human biospecimens that have already been collected by other established programs. Therefore, no consent was obtained for this work. Previously, all human specimens were obtained from the University of Washington (UW) Alzheimer's Disease Research Center after approval by the institutional review board of the Veterans Affairs Puget Sound Health Care System (MIRB# 00331). AD patient diagnosis was confirmed postmortem by neuropathological analysis. Clinically normal subjects were volunteers who were over 65 years of age, never diagnosed with AD, and lacked AD neuropathology at autopsy. AD Brains exhibited Braak stages between V and VI, whereas control brains exhibited Braak stages between I and III. Postmortem frontal lobe tissues were obtained from the middle frontal gyrus tissues that had been rapidly frozen at autopsy (<10 h after death) and stored at −80 °C until use. Demographics of the study sample are listed in Table 1. Hepatocytoma HepG2, glioblastoma U-87 MG and U-118 MG cells (ATCC) were grown in 89% Dulbecco's modified Eagle's medium (DMEM) (Gibco); neuroblastoma SH-SY5Y cells (ATCC) were grown in 89% DMEM with F12 (Gibco). Both media were supplemented with 10% fetal bovine serum (FBS) (Gibco). Glioblastoma LN-229 cells (ATCC) were grown in 94% DMEM supplemented with 5% FBS. All cell cultures were supplemented with 1% penicillin/ streptomycin (Invitrogen) and cultured at 37 °C in a 5% CO_2 atmosphere.

Table 1. Demographics of the PMB study samples.

Subjects	AD	Control
Sample number—n	47	20
Gender—n female (% female)	27 (57.4)	11 (55.0)
APOE ε4+—n (%)	29 (61.7)	3 (15.0)
Age at death—mean (SD)	87.9 (5.9)	88.3 (8.5)
Age at onset—mean (SD)	79.0 (8.0)	N/A
Disease duration—mean years (SD)	9.0 (4.4)	N/A
Postmortem interval—mean hours (SD)	5.0 (2.0)	4.9 (2.3)
CERAD Score		
Absent	0	7
Sparse	0	7
Moderate	11	4
Frequent	36	2
Braak Stage		
I	0	6
II	0	11
III	0	3
IV	0	0
V	15	0
VI	32	0

SD: standard deviation.

2.2. DNA/RNA Extraction and Genotyping

Genomic DNA and RNA were isolated from frozen PMB using the AllPrep DNA/RNA Mini Kit (Qiagen). Nucleic acid concentrations were measured by NanoPhotometer (Implen), and samples were stored at −20 °C prior to use. SNPs (assay #) were genotyped using TaqMan allelic discrimination assays purchased from Thermo Fisher Scientific as follows: rs429358 (C_3084793_20), rs7412 (C_904973_10), rs71352238 (C_98078714_10), rs2075650 (C_3084828_20), rs741780 (C_3084816_10), and rs10119 (C_8711595_10). All procedures were performed according to the manufacturers' protocols. For rs10524523 (poly-T) S/L/VL typing, PCR was performed using primers chr19_50094846F (5′-cctccaaagcattgggatta) and chr19_50095058R (5′-gggacagggaaagaaaacaa). The length of the amplicons was then determined using a QIAxcel Advanced system (Qiagen) based on a high-resolution capillary electrophoresis. The expected amplicon size was calculated to be 179 bp + poly-T length in bp. The observed amplicon size was ≤198 bp for the S variant (poly-T ≤ 19); 199–208 bp for the L variant (poly-T = 20–29); and ≥ 209 bp for the VL variant (poly-T ≥ 30).

2.3. Sequence Alignment and Phylogenetic Tree

The *TOMM40* mRNA and *TOMM40L* transcript reference sequences were obtained from NCBI Nucleotide database. All *TOMM40* pseudogene sequences were extracted from UCSC Genome Browser's UCSC DAS server, using the genomic coordinates (version hg38) obtained from NCBI Genes & Expression's Gene database. The nucleotide sequences were aligned using NIH's BLAST blastn program. The query was optimized for highly similar sequences (megablast). The extracted *TOMM40* pseudogene sequences were blasted against *TOMM40* mRNA reference sequence to query for percent identity. The Phylogenetic Tree was generated from Molecular Data with MEGA (https://doi.org/10.1093.molbev/mst012, accessed on 22 March 2021). The bootstrap value (or node) was calculated from resampling analysis as an indicator of good confidence in specific node. The substitution rate is defined as the number of nucleotides that were substituted per site per unit time.

2.4. Conventional End-Point PCR and Gel Electrophoresis

Expression of pseudogene RNAs was examined by end-point PCR. Total RNAs were extracted from cells and cDNA synthesis was performed using the PrimeScript RT Reagent Kit (Takara Bio, Mountain View, CA, USA). Pseudogene-specific primer sets were used to amplify each pseudogene template or cDNA. Information on the pseudogene-specific primers is listed in Table S1. DNA fragment analysis of the amplification reactions was performed in a QIAxcel (Qiagen).

2.5. Reverse Transcriptase (RT) Reaction and Quantitative PCR (qPCR) Assay

RT-qPCR assays were performed as previously reported [40]. Briefly, a fixed reverse-transcribed cDNA input (5 ng) was amplified using TaqMan assays or SYBR PCR assays in a QuantStudio 5 (Applied Biosystems, Thermo Fisher). The thermal cycling profile consisted of 2 min at 50 °C, 10 min at 95 °C, and then 40 cycles of 15 s at 95 °C and 1 min at 60 °C. The amplification efficiency of both TaqMan and SYBR PCR assays were measured by a standard curve method using serial dilutions in qPCR reactions and calculated using $[10^{(-1/slope)}]$ -1. The calculated amplification efficiency is as follows: 0.92 (total *TOMM40* mRNA); 0.90 (*TOMM40* IVS9); 0.89 (pseudogene P1b/P2); 0.91 (total *TOMM40* Ex4-Ex5). For each sample, qPCR assays were performed in triplicate. Information on primers, probes, and TaqMan assays is listed in Table S1. For *TOMM40* RNA quantification, all reactions were quantified by using a fixed threshold (0.15) in the linear range of amplification and recording the number of cycles (cycle threshold, C_T) required for the fluorescence signal to cross the threshold. To control for the quantity of input RNA, we quantified *ACTB* mRNA as an internal control for each sample and obtained a normalized ΔC_T value: mean of C_T triplicate (target)–mean of the *ACTB* C_T triplicate. In this setting, smaller ΔC_T values indicate higher RNA transcription levels. Additionally, fold change (FC) of *TOMM40* transcription levels of AD to Control subjects was computed as FC (AD) = $2^{-\Delta\Delta Ct}$, where $\Delta\Delta C_T$ = mean ΔC_T (AD)– mean ΔC_T (Control) [41].

2.6. Fraction Estimation of Pseudogene RNA and Surrogate RNA Using Digital PCR (dPCR)

The *P1b/P2* primer set was used to measure levels of pseudogene RNAs and IVS9 primers were used to amplify *TOMM40* surrogate RNA by qPCR. A primer set spanning Ex4 and Ex5 was used for measuring levels of the total *TOMM40* RNA pool. These primers were also used for RT-qPCR (SYBR) assays as listed in Table S1. We performed absolute quantification of RNA levels by QIAcuity dPCR (Qiagen). The QIAcuity carries out fully automated processing including all necessary steps of plate priming, sealing of partitions, thermocycling, and image analysis. We used the the QIAcuity Nanoplate 26K 24-well. For each well, 40 µL reaction contained 13.3 µL of 3x EvaGreen PCR master mix (Qiagen), 0.4 µM of each forward and reverse primer, and a fixed concentration of cDNA template

3 µL (15 ng). The thermal cycling program consisted of 2 min at 95 °C, 40 cycles of 15 s at 95 °C, 20 s at 55 °C, and 1 min at 72 °C, and then 5 min at 40 °C. We computed the fraction of target RNA (pseudogene RNA or surrogate IVS9 RNA) by dividing the number of copies/µL of target RNA by the number of copies/µL of total *TOMM40* RNA pool. For the quality control of QIAcuity dPCR, we replicated the assay with different amounts of template input and showed the reproducibility of the fraction of the target RNA in the total *TOMM40* RNA pool.

2.7. Hydorogen Peroxide Treatment

Twenty-four hours prior to treatment, the cells were seeded at a density of 70–80%. For RNA transcription and mitochondrial DNA (MtDNA) copy number assays, cells were seeded on a 6-well plate, whereas a 96-well plate was used for the mitochondrial membrane potential assay. We searched the literature for the effects of hydrogen peroxide on mitochondrial function. Based on previously published conditions, we tested multiple concentrations (100 µM, 200 µM, 250 µM, 500 µM and 1 mM) of hydrogen peroxide in the cell lines and selected 500 µM as an optimal concentration that maintained good cell viability and had noticeable effects on mitochondrial function. The seeded cells were then treated with 500 µM Hydrogen peroxide, H_2O_2, (Sigma) in growth media. For controls, the same number of cells were plated and cultured without H_2O_2. Cells were collected 24 h post-treatment, subjected to genomic DNA and total RNA isolation, followed by measurement of MtDNA copy number and RNA transcription levels. Three to four independent treatments with H_2O_2 were performed.

2.8. MtDNA Copy Number Assay

Reactions for MtDNA copy number count and single copy reference gene, *HGB* (Hemoglobin), were run separately with 10 ng of DNA in a 384-well optical plate. Each reaction was run in triplicate on QuantStudio 5 (Applied Biosystems, Thermo Fisher, Foster City, CA, USA). The 10 µL reaction included 10 ng of DNA, 5 µL of 2x Power SYBR Green PCR Master Mix (Applied Biosystems, Thermo Fisher), and 0.05 µM of each forward and reverse primer. Thermal cycling profile consisted of 2 min at 50 °C, 10 min at 95 °C, and then 40 cycles of 15 s at 95 °C, 60 s at 56 °C, and 60 s at 72 °C. The ΔC_T method was used to control for the quantity of input DNA for each sample by quantification of *HGB* DNA. The normalized ΔC_T value was calculated: mean of MtDNA C_T triplicate− mean of the *HGB* C_T triplicate. The fold change (FC) of the MtDNA copy number isolated from H_2O_2-treated cells to untreated cells was computed as FC (treated) = $2^{-\Delta\Delta C_T}$, where $\Delta\Delta C_T = \Delta C_T$ (treated) − ΔC_T (untreated) [41].

2.9. Mitochondrial Membrane Potential (MMP) Assay

MMP of the human cell lines was analyzed using a MitoProbe JC-1 assay kit (Thermo Fisher). The cationic dye, JC-1 (5′,6,6′-tetrachloro-1,1′,3,3′-tetraethylbenzimidazolyl-carbocyanine iodide), exhibits potential-dependent accumulation in mitochondria, which is indicated by a fluorescence emission shift from monomeric green (529 nm) to JC-1 aggregates red (590 nm). Consequently, MMP change in response to cellular stimuli is represented by the ratio of red to green fluorescence intensity. The membrane potential disrupter, CCCP (carbonyl cyanide 3-cholorophenylhydrazone), was included in all assays as a control to confirm that the JC-1 response is sensitive to changes in membrane potential. Twenty-four hours prior to the hydrogen peroxide treatment, cells were seeded at a density of 70–80% on a black 96-well plate with a clear bottom. The seeded cells were treated with 500 µM H_2O_2 for 24 h and then assayed for MMP measurements. A quantity of 2 µM JC-1 was added and incubated at 37 °C, 5% CO_2 for 30 min. The reaction plate was washed with PBS and the fluorescence was measured with 488 nm excitation and green (529 nm) or red (590 nm) emission using SpectraMax M2 plate reader (Molecular Devices). All procedures were performed according to the manufacturers' protocols. Fold change (FC) of membrane potential of H_2O_2-treated cells to untreated cells was computed as FC (treated) = ratio of red to green (treated)/ratio of red to green (untreated). Three independent MMP assays were performed for each H_2O_2 treatment for each cell type.

2.10. Statistical Analyses

The qPCR data is expressed as normalized ΔC_T values and was calculated as follows, ΔC_T value = mean of C_T triplicate (target)−mean of the ACTB C_T triplicate. Statistical analyses were performed using independent samples *t*-test using the Statistical Package for the Social Sciences (SPSS) version 19 (SPSS). The MtDNA copy number assay and membrane potential assay were not performed in the same cell setting because the membrane potential assay needs to be conducted on live cells. For this reason, data were not statistically compared.

3. Results

3.1. Presence of Pseudogene RNAs Obscures Accurate Measurement of TOMM40 mRNA

The transcription levels of *TOMM40* mRNA in human brain and its association with AD risk have not been fully established. Studies have reported conflicting results of either up- or down-regulated *TOMM40* mRNA in AD brains. The measurement of a gene's mRNA levels should be a straightforward procedure unless other complications confound this measurement; apparently, an unknown barrier exists in the measurement of *TOMM40* mRNA. To expose this obstacle, we first inspected the specificity of *TOMM40* cDNA by aligning this sequence to the human genome (hg38) using the Blat tool in the UCSC genome browser (http://genome.ucsc.edu/, accessed on 18 March 2021). Besides *TOMM40* itself, the alignment showed six additional hits, including one known gene (*TOMM40L*) with moderate homology to *TOMM40* and five loci with a high degree of sequence homology. All five of these highly homologous loci contain a *TOMM40* cDNA-like sequence lacking either *TOMM40* introns or a full open reading frame, classic characteristics of a pseudogene. When queried the public databases, we found established pseudogene records for four of the five loci in the HUGO gene database (https://www.genenames.org/, accessed on 18 March 2021) with the designated nomenclatures of *TOMM40P1, P2, P3*, and *P4*. We designated the unnamed pseudogene "*TOMM40P1b*" due to its proximity to *TOMM40P1*. The genomic location, span, and similarity to *TOMM40* cDNA of these pseudogenes are listed in Table 2. A phylogenetic tree analysis indicated that they are indeed closely related to each other (Figure 1A).

Because a large portion of the human genome's non-coding regions, including pseudogenes, can produce RNA transcripts [42], we wondered whether these *TOMM40* pseudogenes can be transcribed into RNA. To address this question, we developed new RT-PCR assays designed to specifically amplify each pseudogene's putative RNA transcripts and generated pseudogene-specific DNA templates to serve as positive controls. To generate pseudogene templates, we PCR amplified each pseudogene's genomic region using primer sets (Table S1) flanking immediate up- and down-stream segments of each pseudogene, and then purified these PCR fragments using agarose gel electrophoresis. Each pseudogene template contains the entire pseudogene sequence, thus mimicking its putative cDNA. Except for *TOMM40P1*, which is flanked by heavy repetitive sequences and could not be amplified, we successfully generated DNA templates for the remaining four pseudogenes (*P1b, P2, P3*, and *P4*). Due to high homology between *P1b* and *P2*, as seen in the phylogenetic tree (Figure 1A), we combined these two into a single template of *P1b/P2*. To design specific primers for the *TOMM40* pseudogene RNA assays, we first identified nucleotide variants among sequences of *TOMM40* cDNA and pseudogenes using Clustal Omega's sequence alignment (https://www.ebi.ac.uk/Tools/msa/clustalo/, accessed on 10 June 2019). We then designed allele-specific PCR primers that carry the unique nucleotide(s) of each pseudogene at the 3′-end of the primers (Table S1). We tested these primers' specificities in our collection of pseudogene templates using conventional end-point PCR and capillary gel electrophoresis. The results showed a robust amplification by each specific primer set with its own template, but they also showed various degrees of cross-amplifications with other pseudogene templates (Figure S1), implying that the allele-specific primers cannot fully differentiate each pseudogene. Nevertheless, these primers provide a molecular tool that makes detection of RNA transcripts of the *TOMM40* pseudogenes feasible.

Table 2. Genomic locations and RNA transcripts of *TOMM40*-related genes and pseudogenes.

Gene/Pseudogene	Genomic Sequence (hg38)		RNA Sequence (RefSeq)		
	Coordinate	Span (bp)	Accession #	Size (nt)	BLASTn **
TOMM40	chr19: 44891220-44903689	12,470	NM_001128916	1676	-
TOMM40P1	chr14: 19266948-19268660	1713	NG_022836	1713 *	95.70%
TOMM40P1b	chr14: 19131227-19133057	1831	N/A	1831 *	95.55%
TOMM40P2	chr22: 15853581-15855410	1830	NG_022885	1830 *	95.88%
TOMM40P3	chr5: 3501872-33503327	1456	NG_021878	1456 *	87.26%
TOMM40P4	chr2: 31723017-131724478	1462	NG_023610	1462 *	95.41%
TOMM40L	chr1: 161226060-161230746	4687	NM_032174	2790	70.17%

*: Putative RNA size estimated from corresponding genomic locus. **: % identify when compared to *TOMM40* mRNA.

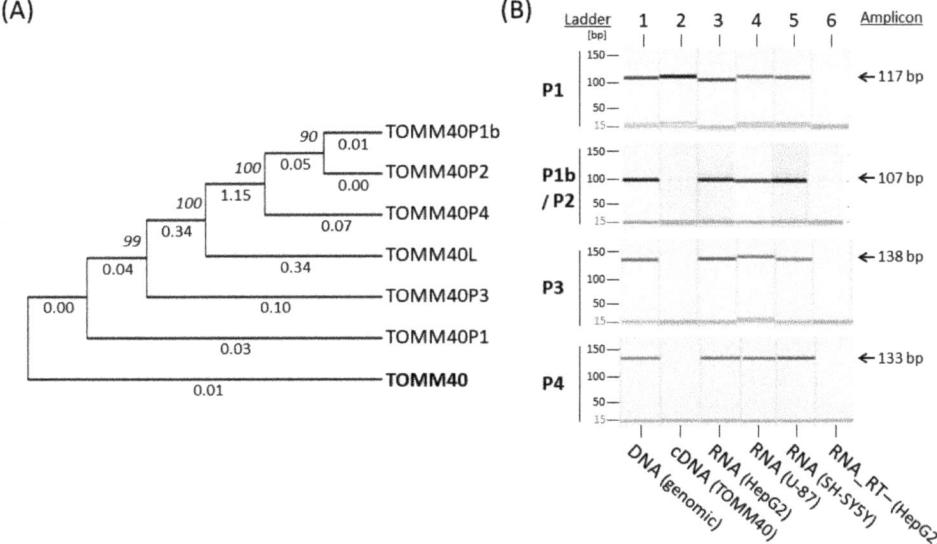

Figure 1. Phylogenetic tree evolution and end-point RT-PCR amplification of *TOMM40* pseudogenes. (**A**) Phylogenetic tree structure of *TOMM40*-related genes and pseudogenes. Numbers above the lines are bootstrap value (or node), and numbers below the lines are substitution rate. (**B**) Capillary gel electrophoresis images of RT-PCR amplified pseudogenes from total RNA of three human cell lines (HepG2, U-87, and SH-SY5Y). Amplicons for pseudogenes *P1*, *P1b*, *P2*, *P3*, and *P4* are shown. Genomic DNA (lane 1) served as a positive control, *TOMM40* cDNA (lane 2) served as a reference for cross-amplification, and RT- RNA (lane 6) served as a negative control.

Using these pseudogene-specific primers, we tested whether putative RNAs of the *TOMM40* pseudogenes could be detected in human cell lines (HepG2, U-87, and SH-SY5Y). Because the expected amplicons of pseudogene RNAs can also be generated from pseudogenes' corresponding genomic DNA, we performed DNase digestion for all RNA isolations and included RT-negative controls for all experiments. In these experiments, we also integrated a *TOMM40* cDNA control that was RT-PCR generated using *TOMM40* cDNA-specific primers. We then performed RT-PCR reactions using total RNA isolated from these cells. The capillary gel electrophoresis showed that all pseudogenes-specific amplicons were amplified in all cell lines tested, with no amplification from RT-negative controls (Figure 1B). None of the pseudogene primer sets amplified *TOMM40* cDNA except the primer set of *P1* (Figure 1B, lane 2), which is likely due to the high homology between *TOMM40* cDNA and *P1*. The similarity of these two sequences in the phylogenetic tree further supported this notion (Figure 1A). Together, these results indicated that all *TOMM40* pseudogenes can produce RNA transcripts and some of these transcripts closely resemble *TOMM40* mRNA.

We also attempted to generate *TOMM40* cDNA-specific primers that could separate *TOMM40* mRNA from the pseudogene RNAs. The *TOMM40* gene consists of 10 exons and six mRNA transcripts (Figure 2A,B). We applied the same allele-specific primer design method to integrate *TOMM40* cDNA-specific nucleotide variants at the 3′-end of the primers. Using this approach, we generated two *TOMM40* cDNA assays that amplify across the splicing junctions of exons 1–2 (Ex1-Ex2) and exons 3–4 (Ex3-Ex4) (Figure 2C). When subjected to RT-PCR experiments, these two assays effectively amplified cDNA templates of *TOMM40*; however, they also showed leaky attributes and cross-amplification of all pseudogene templates (Figure S2). This result suggests that RNA transcripts of *TOMM40* and pseudogenes cannot be fully separated even with the meticulously designed allele-specific primers.

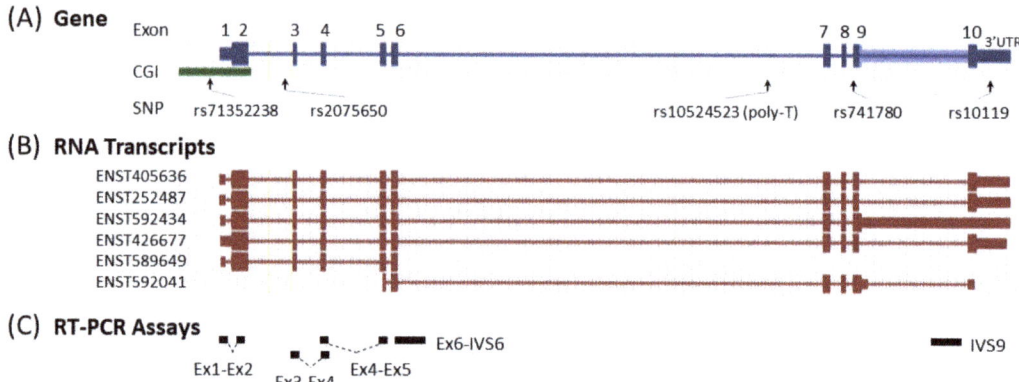

Figure 2. Map of the *TOMM40* gene, RNA transcripts, and RT-PCR assays. (**A**) Structure and position of the exons, CpG island (CGI), and genetic variants (SNP). (**B**) Structure of the six mRNA transcripts defined by the Ensembl database. (**C**) Targets of the RT-PCR assays used in this study.

Learning from these observations, we suspected that the RNA transcription levels determined by commonly used *TOMM40* gene expression assays likely include both RNA species of *TOMM40* and its pseudogenes. We then designed experiments to further examine this possibility. Because exact primer and probe sequences of the commercial assays are not publicly available, we generated a *TOMM40* cDNA assay that spans splice junctions of exons 4-5 (Ex4-Ex5), the general area targeted by a popular *TOMM40* TaqMan expression assay (Thermo Fisher, assay Hs01587378_mH). When tested using end-point PCR, this assay cross-amplified all pseudogene templates (Figure 3A). This result is further evidence that conventional RT-PCR-based *TOMM40* assays likely quantify transcription levels of the entire *TOMM40*-related RNA pool, which consists of both *TOMM40* mRNA and its pseudogenes' RNAs. Such assays cannot provide accurate measurements of true *TOMM40* mRNA.

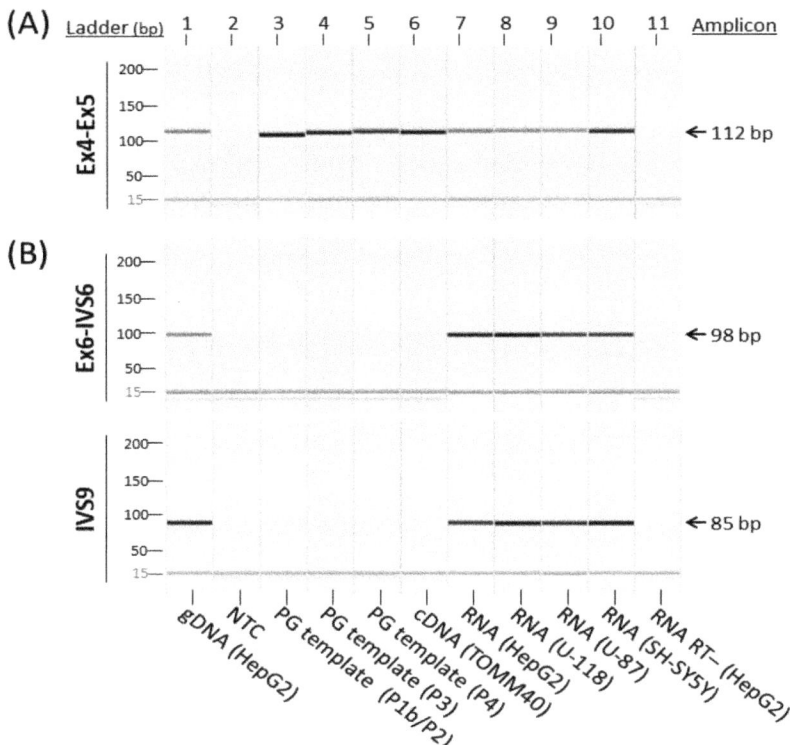

Figure 3. Comparison between a conventional cDNA assay and primary transcript-targeted cDNA assays. Capillary gel electrophoresis images of the RT-PCR amplified *TOMM40* pseudogene amplicons. (**A**) *TOMM40* primer set spans Ex4 and Ex5 that mimics a commercial TaqMan gene expression assay (Thermo Fisher, Waltham, MA, USA, Hs01587378_mH). (**B**) *TOMM40* primer sets targeting primary transcripts of exon 6 to intron 6 (EX6-IVS6) and intron 9 (IVS9). Genomic DNA (lane 1) served as a positive control; no-template control (NTC, lane 2) and RT- RNA (lane 11) served as a negative control.

3.2. Development of TOMM40-Specific RT-PCR Assays

To obtain authentic transcription levels of *TOMM40* mRNA, one could apply a deduction method in which pseudogene transcription levels are subtracted from the total *TOMM40*-related RNA pool. However, we found that most pseudogene-specific primer sets could cross-amplify templates of other pseudogenes (Figure S1), which made the precise measurement of each pseudogene RNA level unfeasible. Instead, we were only able to estimate the fraction of RNA representing pseudogenes within the total *TOMM40* RNA pool. We reasoned that the *P1b/P2* primer set can cross-amplify *P3* and *P4* in addition to its own template (Figure S1); using this one primer set allows us to access the transcription levels representing a large portion of the pseudogene RNAs while excluding amplification of the true *TOMM40* RNA. When they were compared with the results of the Ex4-Ex5 assay, which represents a measurement of the total *TOMM40*-related RNA pool, we were able to approximate the proportion of the pseudogene RNAs. We applied this strategy on total RNA isolated from a subset of PMB tissues (AD $n = 29$ and control $n = 16$) and quantified the cDNA targets using digital PCR (dPCR). From each subject, two amplicons (*P1b/P2* and Ex4–Ex5) were quantified separately, and the absolute quantification count of *P1b/P2* was then divided by the count of Ex4–Ex5 to generate the fraction. The result showed that *P1b/P2* RNA constituted around 10–18% of the total *TOMM40*-related RNA pool (Table S2).

A detailed procedure for this comparison and calculation is listed in the methods section. Provided that this estimation did not include *P1* levels, the actual fraction of the pseudogene RNAs in the total *TOMM40*-related RNA pool could be substantially higher.

The results described above prompted us to conclude that in order to eliminate cross-amplification of *TOMM40* pseudogene RNAs and accurately quantify *TOMM40* transcription, there is a need to develop an unconventional RT-PCR assay. Accordingly, we explored assays designed to target the primary RNA transcript of the *TOMM40*. We reasoned that the main difference between *TOMM40* mRNA and pseudogene RNAs lies in its primary transcript—with pseudogenes lacking intronic sequences. While the primary transcript (pre-mRNA) transcription level is expected to be only a fraction of the spliced mRNA transcription level, this proportion is likely retained consistently across samples obtained from same cell/tissue-types. Thus, an RT-PCR assay based on this principle could provide an accurate surrogate measurement of the actual mRNA transcription level, which can then be used to compare samples from human subjects.

Based on this rationale, we first inspected the RNA structure of *TOMM40* using the Ensembl RNA track in the UCSC genome browser. This track shows the presence of six variants of *TOMM40* RNA transcripts (Figure 2B). To cover the majority of these transcripts, we designed two sets of TaqMan-based assays with one extending from exon 6 into intron 6 (Ex6-IVS6) and a second one extending within intron 9 (IVS9). Map locations of these assays are shown in Figure 2C, and corresponding primers and probe sequences are shown in Table S1. Initial conventional end-point PCR tests with their respective primers showed that both assays amplified RNA samples isolated from all four human cell lines (HepG2, U-118, U-87, and SH-SY5Y) with the expected amplicons. More importantly, no pseudogene amplicons were amplified by these new assays (Figure 3B). We next evaluated these assays in PMB tissues using a TaqMan (primers plus probe) setting in RT-qPCR. When these two assays were compared side by side, they both showed consistent expression patterns ($\approx$2 ΔC_T difference) between AD and control frontal lobes (Figure 4A). Between these two assays, the IVS9 assay showed a higher sensitivity, as indicated by its lower ΔC_T value; thus, we selected this assay for our surrogate quantification of *TOMM40* mRNA. We estimated that RNA levels measured by the IVS9 surrogate assay represent approximately 7–20% of the total *TOMM40*-related RNA pool levels using the same dPCR approach mentioned above (Table S2). We then applied this IVS9 assay to quantify *TOMM40* mRNA transcription levels in human PMB samples and compared these levels to the ones generated from the commercial TaqMan assay (Thermo Fisher, Hs01587378_mH). No differences in *TOMM40* mRNA transcription levels were observed between AD and control when the commercial assay was used. On the contrary, *TOMM40* mRNA showed significantly ($p < 0.001$, *t*-test) higher expression ($\approx$2.5-fold) in AD compared to control when the IVS9 assay was used (Figure 4B).

3.3. Effects of TOMM40 RNA Transcription Levels

With biologically meaningful RNA measurements in hand, we further examined the relationship between *TOMM40* RNA levels and some AD-associated genotypes in human PMB. We first analyzed a set of genetic variants, including the *APOE* ε4-determing SNP rs429358 and five SNPs (rs71352238, rs2075650, rs10524523, rs741780, and rs10119) scattered across *TOMM40* (Figure 2A). For rs10524523, we stratified the "S" and "VL" variants only, excluding the third variant "L", which is linked specifically with the ε4 variant of rs429358 that was analyzed separately. We then tested for associations between stratified alleles and *TOMM40* RNA levels quantified using our IVS9 assay. We observed significant allelic differences in rs10524523 ($p < 0.02$, *t*-test) and in rs741780 ($p < 0.01$, *t*-test; Figure 5). None of the other four SNPs showed any expression associations with their alleles. Additionally, we performed ex vivo experiments to examine the relationship between *TOMM40* RNA levels and selected mitochondria functions. We induced oxidative stress using H_2O_2 in human cell lines (HepG2, U-118, U-87, and LN-229) and compared *TOMM40* RNA levels, MtDNA copy number, and mitochondrial membrane potential. After H_2O_2 treatment, *TOMM40* surrogate RNA levels were increased ($\approx$1.2–1.5-fold higher than the untreated one, which corresponded with a decreased DNA copy number (≈20–60% of the untreated levels) and a decreased membrane potential (≈20–75% of the untreated one; Figure 6). These findings indicate that upregulation of *TOMM40* RNA levels corresponds with mitochondrial dysfunction.

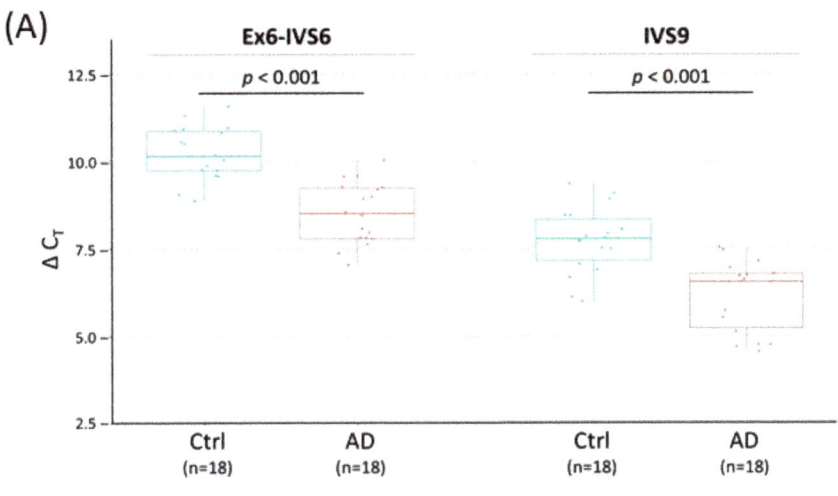

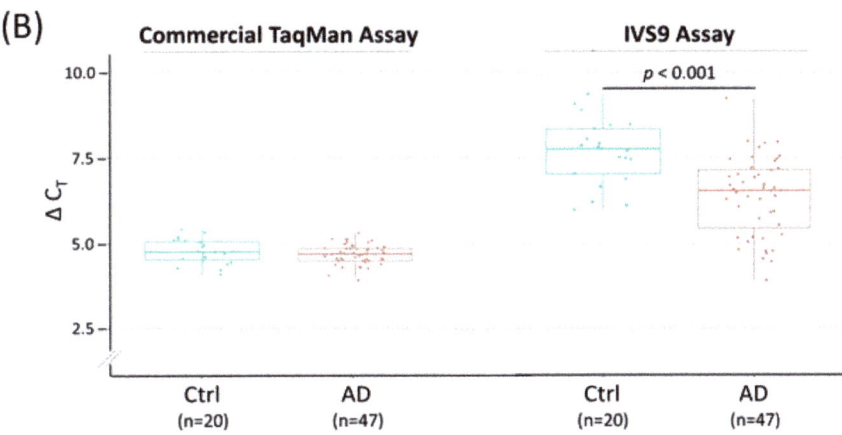

Figure 4. *TOMM40* RNA transcript levels in human PMB tissues by RT-qPCR quantification. Transcription levels of *TOMM40* RNAs are plotted as values of ΔC_T (mean of C_T triplicates (target)—mean of *ACTB* C_T triplicates) and compared between control (Ctrl) (blue) and AD (red) frontal lobes. In this setting, smaller ΔC_T values indicate higher RNA levels. (**A**) Comparison of the two *TOMM40* primary transcript-targeted assays (Ex6-IVS6 vs. IVS9) as the pilot study. (**B**) Comparison of a commercial TaqMan cDNA assay (Thermo Fisher, assay Hs01587378_mH) and the primary transcript-targeted TaqMan IVS9 assay with expanded samples. Numbers in parentheses denote sample size. The *t*-test *p* values are shown where significant differences between Ctrl and AD were detected. Boxplot shows quartiles and median.

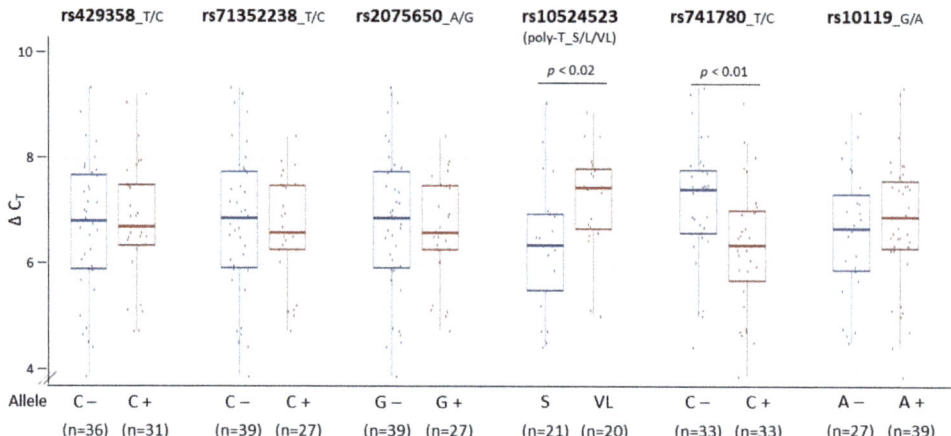

Figure 5. Associations of PMB *TOMM40* RNA levels with genetic variants. Transcription levels of PMB *TOMM40* RNA were measured by the IVS9 assay and compared across six genetic variants. The *t*-test *p* values are shown where significant associations with *TOMM40* intron 6 poly-T SNP rs10524523 ($p < 0.02$) and intron 8 SNP rs741780 ($p < 0.01$) were detected. For rs10524523, the S group includes S/S homozygotes and S/L heterozygotes, and the VL group includes VL/VL homozygotes and VL/L heterozygotes. Numbers in parentheses denote sample size. Boxplot shows quartiles and median.

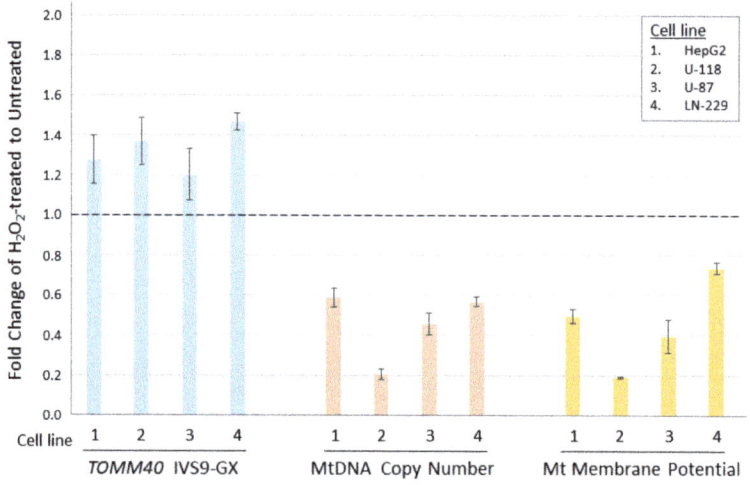

Figure 6. Comparison of *TOMM40* RNA levels with mitochondrial functions in human cell lines. *TOMM40* RNA levels were measured by the IVS9 assay and mitochondrial function assays including MtDNA copy number and Mt membrane potential were performed. The fold change of H_2O_2-treated cells was compared to the untreated cells (set as baseline of 1.0). Graph shows the relationship between the three measurements in response to the oxidative stress. Standard deviation error bars are shown.

4. Discussion

Brain mitochondrial function plays a crucial role in neural plasticity and cognition [43] and is vital to many neural activities. Mitochondrial dysfunction occurs in a variety of psychiatric and neurodegenerative disorders [44], and is a fundamental characteristic of AD [11–14]. The *TOMM40* gene encodes a mitochondrial outer membrane translocase, which plays important roles in importing and sorting proteins for sub-mitochondrial locations. *TOMM40* is an essential gene for mitochondrial maintenance, making it a plausible candidate for influencing AD risk via mitochondrial dysfunction.

The possibility that *TOMM40* plays a direct role in AD risk has always been overshadowed by *APOE*. Located 2.1 kb upstream of *APOE*, the genetic effects of these two genes cannot easily be separated due to the strong LD structure between them [25,45]. Genetic associated signals of *TOMM40* in AD have traditionally been dismissed as surrogate signals of *APOE* [46,47]; however, this viewpoint has gradually shifted to consider *TOMM40* as an independent contributor to AD risk and healthy aging. For example, genetic variants in *TOMM40* have been consistently linked to longevity and healthy aging [48–51]. The SNP rs2075650 located in intron 2 of *TOMM40* has been considered a proxy of the SNP rs429358 that defines the ε4 allele of *APOE* [29]. The G allele of rs2075650 has been associated with a range of phenotypes including reduced longevity [52], reduced BMI [53], and increased low-density lipoprotein cholesterol [54,55], as well as an increased risk of AD [56]. Evidence also suggested that the length of rs10524523 (poly-T) within intron 6 of the *TOMM40* is linked to different levels of risk and age of onset of cognitive decline [57]. Such epidemiological data strongly support the idea that *TOMM40* plays a direct role in cognition and healthy aging. Because age is the most important risk factor for AD, the combined biological consequences of *TOMM40* and *APOE* may represent a molecular mechanism explaining the *APOE* locus' strong genetic association with AD. Additionally, the most recent human genome project has revealed that a large number of functional sites in the genome are *cis*-regulated in nature. Thus, the unique genomic arrangement of the *TOMM40-APOE* gene cluster raises a possibility that genetic variants of the *APOE* locus could relate indirectly to mitochondrial function through LD with *TOMM40*. This concept is also in line with current trending research on the co-regulation of local genes in gene expression through the topological associating domain or 3D genome [58].

When the expression profile of a gene is altered in a disease, it provides credible evidence supporting a direct connection between that gene and the disease. *TOMM40* overexpression at both the transcriptional and translational levels in ovarian cancer has been shown to correlate with increased cell proliferation, migration, and invasion [59,60]. TOM40 protein levels are significantly reduced in brains of Parkinson's disease patients and in α-Syn transgenic mice [61–63]. Significant changes were observed in the mRNA levels of mitochondrial dynamic genes such as fission/fusion-related genes and mitophagy-related genes in blood samples of AD patients [39,64]. However, the altered expression profiles (both mRNA and protein) of *TOMM40* in AD have not been clearly established. Whether the expression level of *TOMM40* is up- or down-regulated in AD brains has not been consistently observed. In one study, *TOMM40* mRNA was shown to be downregulated in 6 of 14 AD frontal lobes but upregulated in the remaining eight [39]. Such conflicting results suggest that the quantification of *TOMM40* mRNA may not be straightforward and may be complicated by other biological processes.

To determine the source of inconsistent *TOMM40* transcription levels observed in the human brain, we revisited the fundamental basics by examining the specificity of the *TOMM40* cDNA sequence. We were surprised to find that human genome carries five *TOMM40*-related pseudogenes, which all share high homology (87–96%) with *TOMM40* cDNA. We were even more surprised to find that all five pseudogenes produce RNA transcripts in various human cell lines and PMB tissues. Currently, the biological functions and/or consequences of these *TOMM40* pseudogene RNAs are undefined. The *TOMM40* pseudogenes are scattered across the human genome with unique flanking DNA sequences; thus, each of them is likely independently regulated. It is plausible that some of these RNAs can serve as templates to produce small peptides, but, to our knowledge, this has never been investigated. Another potential function of these *TOMM40* pseudogene RNAs is to regulate the transcription of *TOMM40*. Carrying highly similar sequences, these RNAs could compete with *TOMM40* mRNA for binding to proteins in transcription or post-transcription machineries. Potentially, these pseudogene RNAs could provide an RNA buffer in response to various stress/stimuli, and this buffer effect might minimize severe fluctuations in

TOMM40 RNA production. Indirect evidence supporting this concept comes from our experiment of comparing TOMM40 cDNA assays. The commercial TOMM40 assay, which measures both TOMM40 mRNA and pseudogene RNAs, has much tighter transcription levels across the PMB samples when compared to the IVS9 assay that specifically measures TOMM40 RNA only. One explanation is that this commercial assay congests all the TOMM40-related RNAs together, which provide a cushion effect to reduce the variability of single RNA measurement. The estimated fraction of pseudogene RNAs in the total TOMM40-related RNA pool is approximately 10–18% using a single pseudogene assay (P1b/P2) that cross-amplified four out of five pseudogenes. If the transcription levels of all five pseudogenes can be precisely measured, the overall fraction is likely to be significantly higher than this estimation. These results raise an interesting question: why does TOMM40 gene expression need to be rigorously regulated or guarded by such a complex system from a whole genome setting?

Due to the cross-amplification of pseudogene RNAs, the conventional RT-PCR assays cannot provide an accurate measurement of TOMM40 mRNA. This challenge prompted us to develop an alternative approach, which targets the primary RNA transcript of TOMM40 and eliminates undesired co-measurement of pseudogene RNAs. The major difference between the primary transcript and mRNA lies in the RNA splicing. The splicing efficiency depends on the splicing kinetics, transcriptional and splicing regulators, transcription rate, intron length, exon position, RNA structure, and chromatin signatures, including histone marks and DNA methylation [65,66]. It has been shown that the splicing efficiency of pre-mRNA varies greatly across genes [67,68]. Due to the splicing process, actual transcription levels are not the same between the primary transcript and spliced mRNA. However, the level of primary transcript can provide a surrogate measurement for mRNA. Surrogate TOMM40 transcription levels, which were measured by the primary transcript-targeted assays (Ex6-IVS6 and IVS9), had a similar profile with ≈ 2 ΔC_T value separating AD and control PMB samples. This result indicated that both measurements were consistent across samples and were suitable to serve as surrogate measurements of TOMM40 mRNA. Between the two assays, the IVS9 has higher transcription levels (lower ΔC_T value) when compared to Ex6-IVS6. This difference could be due to either the slower splicing kinetics of IVS9, or simply the presence of a TOMM40 mRNA transcript (ENST592434) that retains the entire intron 9 in its mRNA structure.

After resolving the complication of pseudogene RNAs co-measurement, we demonstrated that the surrogate transcription level of TOMM40 RNA is roughly 2.5-fold higher in AD compared to the control frontal lobe. Our results are opposite to prior published studies showing that TOMM40 RNA is downregulated in AD blood [35–38]. Although this opposition could reflect different TOMM40 regulatory pathways between blood and the CNS, it also suggested that the transcription levels of TOMM40 pseudogene RNAs could vary across different tissues. Increased TOMM40 RNA transcription in AD brains could be a consequence of prolonged mitochondrial dysfunction, which triggers a feedback response to upregulate structural proteins (e.g., TOM40) to compensate for the compromised mitochondrial function. The upregulated TOMM40 RNA in AD brains has the same trend as the upregulation of APOE mRNA in AD brains compared to control [40]. The consistent upregulation of both TOMM40 and APOE in AD brains makes the concept of co-regulation of these genes through the same topological associating domain even more appealing. Whether the upregulation of TOMM40 RNA is truly associated with AD risk remains to be further validated with a larger sample size and across different brain regions. Nevertheless, this study provides a new molecular tool for measuring TOMM40 RNA, making future expanded studies feasible.

Our genetic association analyses revealed no difference between TOMM40 RNA transcription levels with either APOE rs429358 variants (C/ε4+ vs. C/ε4−) or TOMM40 rs2075650 variants (G+ vs. G−), two SNPs that have been consistently linked to AD risk through GWAS studies. This lack of association suggested that these two genetic variants may not have a direct impact on the increased TOMM40 RNA levels in AD.

On the contrary, the SNPs rs10527523 and rs741780 showed significant allelic differences associated with *TOMM40* RNA transcription. Because these two SNPs are located between introns 6 and 8 of the *TOMM40* gene, this region of *TOMM40* might contain functional regulatory elements that influence the transcription of *TOMM40*. In the case of rs10524523, we observed a higher transcription level (lower ΔC_T value) of *TOMM40* RNA in the "S" variant when compared to the "VL" allele. This result is opposite to the study of Linnertz et al., who showed *TOMM40* mRNA levels were lower in "S" homozygotes compared with "VL" homozygotes in the AD brain [24]. Again, the inclusion or exclusion of *TOMM40* pseudogene RNAs transcription levels could account for these conflicting results.

Mitochondria are involved in several cellular functions and are essential for energy production; they are the main organelles that provide energy for brain cells. Indeed, neurons are particularly sensitive to changes in mitochondrial function [69], and mitochondrial injury can have severe consequences for neuronal function and survival [70]. We studied two mitochondrial function-related phenotypes (copy number and membrane potential) and their associations with *TOMM40* RNA transcription levels in human cell lines. MtDNA copy number is a measure of the number of mitochondrial genomes per cell and is a proxy for mitochondrial function [71–73]. Significant differences in this copy number have been reported across different brain regions, and these variations were more pronounced in patients affected by neurodegenerative disorders [74]. Studies have also shown that MtDNA levels were decreased by 30–50% in the frontal cortex of AD patients when compared to controls [75,76]. Mitochondrial membrane potential, which is used by ATP synthase to make ATP, serves as an intermediate form of energy storage for cells. Normally, cells maintain stable levels of mitochondrial membrane potential to carry out various cellular functions [77–79]. This membrane potential is altered due to physiological activity on a transient basis, but a prolonged alteration could compromise the viability of the cells and cause irreversible damage [80]. Our analyses showed that the increased *TOMM40* RNA levels are associated with a lower MtDNA copy number and a lower mitochondrial membrane potential, which together signified a decrease in mitochondrial function. This observation could be explained by a feedback response to restore mitochondria function via upregulation of *TOMM40* mRNA. As a translocase of outer mitochondrial membrane [9,10], TOM40 protein plays a role in importing proteins for the assembly of the mitochondrial inner membrane respiratory chain and mitochondrial matrix proteins involved in oxidative respiration. Increased *TOMM40* RNA transcription associated with AD could lead to changes in mitochondrial protein import, which might affect maintenance of mitochondrial membrane potential and overall mitochondrial function.

5. Conclusions

Here, we developed a novel assay to measure true *TOMM40* RNA transcription levels with high specificity and sensitivity, circumventing the unintentional co-measurement of *TOMM40* pseudogene RNAs. This assay enabled us to accurately investigate the RNA transcription profile of *TOMM40* associated with AD. The PMB work showed that *TOMM40* mRNA is upregulated in AD vs. control frontal lobe. The ex vivo cultured cell line work showed that upregulation of *TOMM40* RNA is likely associated with compromised mitochondrial function. Although future work using a larger sample size is needed to replicate these results, this work pioneers a valuable blueprint to assess *APOE*-independent effects of *TOMM40* in AD risk. Our findings define a new paradigm of *TOMM40* gene regulation and provide novel insight into the transcriptional pathway of *TOMM40*. This pathway involves not only the production of multiple *TOMM40* mRNA species, but also a pseudogene-imparted transcriptional program. Many epidemiology studies strongly support the idea that *TOMM40* contributes to healthy aging. Because age is the most important known risk factor for AD, it raises an interesting question: could the incidence of AD be a byproduct of a compromised longevity pathway that is carefully guarded via *TOMM40*-imparted mitochondrial function?

Supplementary Materials: The following is available online at https://www.mdpi.com/article/10.3390/genes12060871/s1, Figure S1. Allele-specific primers of *TOMM40* pseudogenes can cross-amplify other pseudogenes. Figure S2: Conventional *TOMM40* cDNA assays can cross-amplify *TOMM40* pseudogenes. Table S1. Primers, probes, and TaqMan assays. Table S2. Fractions of *TOMM40* pseudogene *P1b/P2* RNAs and IVS9 RNA in total *TOMM40*-related RNA pool, measured by dPCR.

Author Contributions: Conceptualization, E.-G.L. and C.-E.Y.; Methodology, E.-G.L., C.-E.Y., L.L. and S.C.; Formal analysis, E.-G.L., S.C. and J.T.; Resources, E.-G.L., J.T. and L.L.; Data curation, E.-G.L. and S.C.; Writing—original draft preparation, C.-E.Y.; Writing—review and editing, E.-G.L. and J.T.; Visualization, E.-G.L., C.-E.Y. and S.C.; Supervision, E.-G.L. and C.-E.Y.; Project administration, C.-E.Y.; Funding acquisition, C.-E.Y. All authors have read and agreed to the published version of the manuscript.

Funding: This research was funded by Merit Review Awards, BX000933 and BX004823, from the U.S. Department of Veterans Affairs Office of Research and Development Biomedical Laboratory Research Program. The contents do not represent the views of the U.S. Department of Veterans Affairs or the United States Government.

Institutional Review Board Statement: Ethical review and approval were waived for this study, due to that this work used deidentified human biospecimens that have already been collected by other established programs. Previously, all human specimens were obtained from the University of Washington (UW) Alzheimer's Disease Research Center after approval by the institutional review board of the Veterans Affairs Puget Sound Health Care System (MIRB# 00331).

Informed Consent Statement: Patient consent was waived due to that this work used deidentified human biospecimens that have already been collected by other established programs.

Data Availability Statement: All relevant data are within the manuscript and its Supporting Information files.

Conflicts of Interest: The authors declare no conflict of interest.

References

1. Takei, N.; Miyashita, A.; Tsukie, T.; Arai, H.; Asada, T.; Imagawa, M.; Shoji, M.; Higuchi, S.; Urakami, K.; Kimura, H.; et al. Genetic association study on in and around the APOE in late-onset Alzheimer disease in Japanese. *Genomics* **2009**, *93*, 441–448. [CrossRef] [PubMed]
2. Lambert, J.C.; Ibrahim-Verbaas, C.A.; Harold, D.; Naj, A.C.; Sims, R.; Bellenguez, C.; DeStafano, A.L.; Bis, J.C.; Beecham, G.W.; Grenier-Boley, B.; et al. Meta-analysis of 74,046 individuals identifies 11 new susceptibility loci for Alzheimer's disease. *Nat. Genet.* **2013**, *45*, 1452–1458. [CrossRef]
3. Beecham, G.W.; Hamilton, K.; Naj, A.C.; Martin, E.R.; Huentelman, M.; Myers, A.J.; Corneveaux, J.J.; Hardy, J.; Vonsattel, J.P.; Younkin, S.G.; et al. Genome-wide association meta-analysis of neuropathologic features of Alzheimer's disease and related dementias. *PLoS Genet.* **2014**, *10*, e1004606. [CrossRef] [PubMed]
4. Yashin, A.I.; Fang, F.; Kovtun, M.; Wu, D.; Duan, M.; Arbeev, K.; Akushevich, I.; Kulminski, A.; Culminskaya, I.; Zhbannikov, I.; et al. Hidden heterogeneity in Alzheimer's disease: Insights from genetic association studies and other analyses. *Exp. Gerontol.* **2018**, *107*, 148–160. [CrossRef] [PubMed]
5. Sillen, A.; Forsell, C.; Lilius, L.; Axelman, K.; Bjork, B.F.; Onkamo, P.; Kere, J.; Winblad, B.; Graff, C. Genome scan on Swedish Alzheimer's disease families. *Mol. Psychiatry* **2005**, *11*, 182–186. [CrossRef]
6. Hendrie, H.C.; Murrell, J.; Baiyewu, O.; Lane, K.A.; Purnell, C.; Ogunniyi, A.; Unverzagt, F.W.; Hall, K.; Callahan, C.M.; Saykin, A.J.; et al. APOE epsilon4 and the risk for Alzheimer disease and cognitive decline in African Americans and Yoruba. *Int. Psychogeriatr.* **2014**, *26*, 977–985. [CrossRef]
7. Rajabli, F.; Feliciano, B.E.; Celis, K.; Hamilton-Nelson, K.L.; Whitehead, P.L.; Adams, L.D.; Bussies, P.L.; Manrique, C.P.; Rodriguez, A.; Rodriguez, V.; et al. Ancestral origin of ApoE epsilon4 Alzheimer disease risk in Puerto Rican and African American populations. *PLoS Genet.* **2018**, *14*, e1007791. [CrossRef]
8. Morris, J.C.; Schindler, S.E.; McCue, L.M.; Moulder, K.L.; Benzinger, T.L.S.; Cruchaga, C.; Fagan, A.M.; Grant, E.; Gordon, B.A.; Holtzman, D.M.; et al. Assessment of Racial Disparities in Biomarkers for Alzheimer Disease. *JAMA Neurol.* **2019**. [CrossRef]
9. Humphries, A.D.; Streimann, I.C.; Stojanovski, D.; Johnston, A.J.; Yano, M.; Hoogenraad, N.J.; Ryan, M.T. Dissection of the Mitochondrial Import and Assembly Pathway for Human Tom40. *J. Biol. Chem.* **2005**, *280*, 11535–11543. [CrossRef]
10. Hoogenraad, N.J.; Ward, L.A.; Ryan, M.T. Import and assembly of proteins into mitochondria of mammalian cells. *Biochim. Biophys. Acta* **2002**, *1592*, 97–105. [CrossRef]

11. Swerdlow, R.H.; Burns, J.M.; Khan, S.M. The Alzheimer's disease mitochondrial cascade hypothesis: Progress and perspectives. *Biochim. Biophys. Acta* **2014**, *1842*, 1219–1231. [CrossRef] [PubMed]
12. Kwong, J.Q.; Beal, M.F.; Manfredi, G. The role of mitochondria in inherited neurodegenerative diseases. *J. Neurochem.* **2006**, *97*, 1659–1675. [CrossRef]
13. Kapogiannis, D.; Mattson, M.P. Disrupted energy metabolism and neuronal circuit dysfunction in cognitive impairment and Alzheimer's disease. *Lancet Neurol.* **2011**, *10*, 187–198. [CrossRef]
14. Agrawal, I.; Jha, S. Mitochondrial Dysfunction and Alzheimer's Disease: Role of Microglia. *Front. Aging Neurosci.* **2020**, *12*, 252. [CrossRef]
15. Hirai, K.; Aliev, G.; Nunomura, A.; Fujioka, H.; Russell, R.L.; Atwood, C.S.; Johnson, A.B.; Kress, Y.; Vinters, H.V.; Tabaton, M.; et al. Mitochondrial abnormalities in Alzheimer's disease. *J. Neurosci.* **2001**, *21*, 3017–3023. [CrossRef]
16. Zhu, X.; Perry, G.; Moreira, P.I.; Aliev, G.; Cash, A.D.; Hirai, K.; Smith, M.A. Mitochondrial abnormalities and oxidative imbalance in Alzheimer disease. *J. Alzheimers Dis.* **2006**, *9*, 147–153. [CrossRef]
17. Reddy, P.H.; Beal, M.F. Amyloid beta, mitochondrial dysfunction and synaptic damage: Implications for cognitive decline in aging and Alzheimer's disease. *Trends Mol. Med.* **2008**, *14*, 45–53. [CrossRef]
18. Wang, X.; Su, B.; Zheng, L.; Perry, G.; Smith, M.A.; Zhu, X. The role of abnormal mitochondrial dynamics in the pathogenesis of Alzheimer's disease. *J. Neurochem.* **2009**, *109*, 153–159. [CrossRef] [PubMed]
19. Zhu, Z.; Yang, Y.; Xiao, Z.; Zhao, Q.; Wu, W.; Liang, X.; Luo, J.; Cao, Y.; Shao, M.; Guo, Q.; et al. TOMM40 and APOE variants synergistically increase the risk of Alzheimer's disease in a Chinese population. *Aging Clin. Exp. Res.* **2020**. [CrossRef]
20. Roses, A.D. An inherited variable poly-T repeat genotype in TOMM40 in Alzheimer disease. *Arch. Neurol.* **2010**, *67*, 536–541. [CrossRef]
21. Lutz, M.W.; Crenshaw, D.G.; Saunders, A.M.; Roses, A.D. Genetic variation at a single locus and age of onset for Alzheimer's disease. *Alzheimers Dement.* **2010**, *6*, 125–131. [CrossRef] [PubMed]
22. Lutz, M.W.; Sundseth, S.S.; Burns, D.K.; Saunders, A.M.; Hayden, K.M.; Burke, J.R.; Welsh-Bohmer, K.A.; Roses, A.D. A Genetics-based Biomarker Risk Algorithm for Predicting Risk of Alzheimer's Disease. *Alzheimers Dement.* **2016**, *2*, 30–44. [CrossRef] [PubMed]
23. Crenshaw, D.G.; Gottschalk, W.K.; Lutz, M.W.; Grossman, I.; Saunders, A.M.; Burke, J.R.; Welsh-Bohmer, K.A.; Brannan, S.K.; Burns, D.K.; Roses, A.D. Using genetics to enable studies on the prevention of Alzheimer's disease. *Clin. Pharm.* **2013**, *93*, 177–185. [CrossRef]
24. Linnertz, C.; Anderson, L.; Gottschalk, W.; Crenshaw, D.; Lutz, M.W.; Allen, J.; Saith, S.; Mihovilovic, M.; Burke, J.R.; Welsh-Bohmer, K.A.; et al. The cis-regulatory effect of an Alzheimer's disease-associated poly-T locus on expression of TOMM40 and apolipoprotein E genes. *Alzheimers Dement.* **2014**, *10*, 541–551. [CrossRef]
25. Yu, C.E.; Seltman, H.; Peskind, E.R.; Galloway, N.; Zhou, P.X.; Rosenthal, E.; Wijsman, E.M.; Tsuang, D.W.; Devlin, B.; Schellenberg, G.D. Comprehensive analysis of APOE and selected proximate markers for late-onset Alzheimer's disease: Patterns of linkage disequilibrium and disease/marker association. *Genomics* **2007**, *89*, 655–665. [CrossRef] [PubMed]
26. Blue, E.E.; Cheng, A.; Chen, S.; Yu, C.E.; Alzheimer's Disease Genetics, C. Association of Uncommon, Noncoding Variants in the APOE Region With Risk of Alzheimer Disease in Adults of European Ancestry. *JAMA Netw. Open* **2020**, *3*, e2017666. [CrossRef]
27. Bekris, L.M.; Millard, S.P.; Galloway, N.M.; Vuletic, S.; Albers, J.J.; Li, G.; Galasko, D.R.; DeCarli, C.; Farlow, M.R.; Clark, C.M.; et al. Multiple SNPs within and surrounding the apolipoprotein E gene influence cerebrospinal fluid apolipoprotein E protein levels. *J. Alzheimers Dis.* **2008**, *13*, 255–266. [CrossRef] [PubMed]
28. Bekris, L.M.; Galloway, N.M.; Montine, T.J.; Schellenberg, G.D.; Yu, C.E. APOE mRNA and protein expression in postmortem brain are modulated by an extended haplotype structure. *Am. J. Med. Genet. B Neuropsychiatr. Genet.* **2009**. [CrossRef]
29. Bekris, L.M.; Lutz, F.; Yu, C.-E. Functional analysis of APOE locus genetic variation implicates regional enhancers in the regulation of both TOMM40 and APOE. *J. Hum. Genet.* **2012**, *57*, 18–25. [CrossRef]
30. Hansson Petersen, C.A.; Alikhani, N.; Behbahani, H.; Wiehager, B.; Pavlov, P.F.; Alafuzoff, I.; Leinonen, V.; Ito, A.; Winblad, B.; Glaser, E.; et al. The amyloid beta-peptide is imported into mitochondria via the TOM import machinery and localized to mitochondrial cristae. *Proc. Natl. Acad. Sci. USA* **2008**, *105*, 13145–13150. [CrossRef]
31. Anandatheerthavarada, H.K.; Devi, L. Mitochondrial translocation of amyloid precursor protein and its cleaved products: Relevance to mitochondrial dysfunction in Alzheimer's disease. *Rev. Neurosci.* **2007**, *18*, 343–354. [CrossRef] [PubMed]
32. Cenini, G.; Rub, C.; Bruderek, M.; Voos, W. Amyloid beta-peptides interfere with mitochondrial preprotein import competence by a coaggregation process. *Mol. Biol. Cell* **2016**, *27*, 3257–3272. [CrossRef] [PubMed]
33. Schaefer, P.M.; von Einem, B.; Walther, P.; Calzia, E.; von Arnim, C.A. Metabolic Characterization of Intact Cells Reveals Intracellular Amyloid Beta but Not Its Precursor Protein to Reduce Mitochondrial Respiration. *PLoS ONE* **2016**, *11*, e0168157. [CrossRef]
34. Devi, L.; Prabhu, B.M.; Galati, D.F.; Avadhani, N.G.; Anandatheerthavarada, H.K. Accumulation of amyloid precursor protein in the mitochondrial import channels of human Alzheimer's disease brain is associated with mitochondrial dysfunction. *J. Neurosci.* **2006**, *26*, 9057–9068. [CrossRef] [PubMed]
35. Mise, A.; Yoshino, Y.; Yamazaki, K.; Ozaki, Y.; Sao, T.; Yoshida, T.; Mori, T.; Mori, Y.; Ochi, S.; Iga, J.I.; et al. TOMM40 and APOE Gene Expression and Cognitive Decline in Japanese Alzheimer's Disease Subjects. *J. Alzheimers Dis.* **2017**, *60*, 1107–1117. [CrossRef]

36. Chong, M.S.; Goh, L.K.; Lim, W.S.; Chan, M.; Tay, L.; Chen, G.; Feng, L.; Ng, T.P.; Tan, C.H.; Lee, T.S. Gene expression profiling of peripheral blood leukocytes shows consistent longitudinal downregulation of TOMM40 and upregulation of KIR2DL5A, PLOD1, and SLC2A8 among fast progressors in early Alzheimer's disease. *J. Alzheimers Dis.* **2013**, *34*, 399–405. [CrossRef]
37. Lee, T.S.; Goh, L.; Chong, M.S.; Chua, S.M.; Chen, G.B.; Feng, L.; Lim, W.S.; Chan, M.; Ng, T.P.; Krishnan, K.R. Downregulation of TOMM40 expression in the blood of Alzheimer disease subjects compared with matched controls. *J. Psychiatr Res.* **2012**, *46*, 828–830. [CrossRef]
38. Goh, L.K.; Lim, W.S.; Teo, S.; Vijayaraghavan, A.; Chan, M.; Tay, L.; Ng, T.P.; Tan, C.H.; Lee, T.S.; Chong, M.S. TOMM40 alterations in Alzheimer's disease over a 2-year follow-up period. *J. Alzheimers Dis.* **2015**, *44*, 57–61. [CrossRef]
39. Manczak, M.; Calkins, M.J.; Reddy, P.H. Impaired mitochondrial dynamics and abnormal interaction of amyloid beta with mitochondrial protein Drp1 in neurons from patients with Alzheimer's disease: Implications for neuronal damage. *Hum. Mol. Genet.* **2011**, *20*, 2495–2509. [CrossRef]
40. Lee, E.G.; Tulloch, J.; Chen, S.; Leong, L.; Saxton, A.D.; Kraemer, B.; Darvas, M.; Keene, C.D.; Shutes-David, A.; Todd, K.; et al. Redefining transcriptional regulation of the APOE gene and its association with Alzheimer's disease. *PLoS ONE* **2020**, *15*, e0227667. [CrossRef]
41. Livak, K.J.; Schmittgen, T.D. Analysis of relative gene expression data using real-time quantitative PCR and the 2(-Delta Delta C(T)) Method. *Methods* **2001**, *25*, 402–408. [CrossRef]
42. Witek, J.; Mohiuddin, S.S. Biochemistry, Pseudogenes. In *StatPearls*; StatPearls Publishing: Treasure Island, FL, USA, 2021.
43. Cheng, A.; Hou, Y.; Mattson, M.P. Mitochondria and neuroplasticity. *ASN Neuro* **2010**, *2*, e00045. [CrossRef]
44. Markham, A.; Bains, R.; Franklin, P.; Spedding, M. Changes in mitochondrial function are pivotal in neurodegenerative and psychiatric disorders: How important is BDNF? *Br. J. Pharm.* **2014**, *171*, 2206–2229. [CrossRef] [PubMed]
45. Roses, A.; Sundseth, S.; Saunders, A.; Gottschalk, W.; Burns, D.; Lutz, M. Understanding the genetics of APOE and TOMM40 and role of mitochondrial structure and function in clinical pharmacology of Alzheimer's disease. *Alzheimers Dement.* **2016**, *12*, 687–694. [CrossRef] [PubMed]
46. Martin, E.R.; Lai, E.H.; Gilbert, J.R.; Rogala, A.R.; Afshari, A.J.; Riley, J.; Finch, K.L.; Stevens, J.F.; Livak, K.J.; Slotterbeck, B.D.; et al. SNPing away at complex diseases: Analysis of single-nucleotide polymorphisms around APOE in Alzheimer disease. *Am. J. Hum. Genet.* **2000**, *67*, 383–394. [CrossRef]
47. Martin, E.R.; Gilbert, J.R.; Lai, E.H.; Riley, J.; Rogala, A.R.; Slotterbeck, B.D.; Sipe, C.A.; Grubber, J.M.; Warren, L.L.; Conneally, P.M.; et al. Analysis of association at single nucleotide polymorphisms in the APOE region. *Genomics* **2000**, *63*, 7–12. [CrossRef]
48. Zhang, C.; Pierce, B.L. Genetic susceptibility to accelerated cognitive decline in the US Health and Retirement Study. *Neurobiol. Aging* **2014**, *35*, 1512.e11–1512.e18. [CrossRef]
49. Davies, G.; Armstrong, N.; Bis, J.C.; Bressler, J.; Chouraki, V.; Giddaluru, S.; Hofer, E.; Ibrahim-Verbaas, C.A.; Kirin, M.; Lahti, J.; et al. Genetic contributions to variation in general cognitive function: A meta-analysis of genome-wide association studies in the CHARGE consortium (N = 53949). *Mol. Psychiatry* **2015**, *20*, 183–192. [CrossRef]
50. Fortney, K.; Dobriban, E.; Garagnani, P.; Pirazzini, C.; Monti, D.; Mari, D.; Atzmon, G.; Barzilai, N.; Franceschi, C.; Owen, A.B.; et al. Genome-Wide Scan Informed by Age-Related Disease Identifies Loci for Exceptional Human Longevity. *PLoS Genet.* **2015**, *11*, e1005728. [CrossRef]
51. Liu, X.; Song, Z.; Li, Y.; Yao, Y.; Fang, M.; Bai, C.; An, P.; Chen, H.; Chen, Z.; Tang, B.; et al. Integrated genetic analyses revealed novel human longevity loci and reduced risks of multiple diseases in a cohort study of 15,651 Chinese individuals. *Aging Cell* **2021**, *20*, e13323. [CrossRef] [PubMed]
52. Deelen, J.; Beekman, M.; Uh, H.W.; Helmer, Q.; Kuningas, M.; Christiansen, L.; Kremer, D.; van der Breggen, R.; Suchiman, H.E.; Lakenberg, N.; et al. Genome-wide association study identifies a single major locus contributing to survival into old age; the APOE locus revisited. *Aging Cell* **2011**, *10*, 686–698. [CrossRef]
53. Guo, Y.; Lanktree, M.B.; Taylor, K.C.; Hakonarson, H.; Lange, L.A.; Keating, B.J.; Consortium, I.K.S. Gene-centric meta-analyses of 108 912 individuals confirm known body mass index loci and reveal three novel signals. *Hum. Mol. Genet.* **2013**, *22*, 184–201. [CrossRef]
54. Middelberg, R.P.; Ferreira, M.A.; Henders, A.K.; Heath, A.C.; Madden, P.A.; Montgomery, G.W.; Martin, N.G.; Whitfield, J.B. Genetic variants in LPL, OASL and TOMM40/APOE-C1-C2-C4 genes are associated with multiple cardiovascular-related traits. *BMC Med. Genet.* **2011**, *12*, 123. [CrossRef]
55. Sandhu, M.S.; Waterworth, D.M.; Debenham, S.L.; Wheeler, E.; Papadakis, K.; Zhao, J.H.; Song, K.; Yuan, X.; Johnson, T.; Ashford, S.; et al. LDL-cholesterol concentrations: A genome-wide association study. *Lancet* **2008**, *371*, 483–491. [CrossRef]
56. Denny, J.C.; Bastarache, L.; Ritchie, M.D.; Carroll, R.J.; Zink, R.; Mosley, J.D.; Field, J.R.; Pulley, J.M.; Ramirez, A.H.; Bowton, E.; et al. Systematic comparison of phenome-wide association study of electronic medical record data and genome-wide association study data. *Nat. Biotechnol.* **2013**, *31*, 1102–1110. [CrossRef] [PubMed]
57. Roses, A.D.; Lutz, M.W.; Crenshaw, D.G.; Grossman, I.; Saunders, A.M.; Gottschalk, W.K. TOMM40 and APOE: Requirements for replication studies of association with age of disease onset and enrichment of a clinical trial. *Alzheimers Dement.* **2013**, *9*, 132–136. [CrossRef]
58. Zheng, H.; Xie, W. The role of 3D genome organization in development and cell differentiation. *Nat. Rev. Mol. Cell Biol.* **2019**, *20*, 535–550. [CrossRef] [PubMed]
59. Leek, B.F. Abdominal and pelvic visceral receptors. *Br. Med. Bull.* **1977**, *33*, 163–168. [CrossRef]

60. Dowling, P.; Meleady, P.; Dowd, A.; Henry, M.; Glynn, S.; Clynes, M. Proteomic analysis of isolated membrane fractions from superinvasive cancer cells. *Biochim. Biophys. Acta* **2007**, *1774*, 93–101. [CrossRef]
61. Heinemeyer, T.; Stemmet, M.; Bardien, S.; Neethling, A. Underappreciated Roles of the Translocase of the Outer and Inner Mitochondrial Membrane Protein Complexes in Human Disease. *DNA Cell Biol.* **2019**, *38*, 23–40. [CrossRef] [PubMed]
62. Bender, A.; Desplats, P.; Spencer, B.; Rockenstein, E.; Adame, A.; Elstner, M.; Laub, C.; Mueller, S.; Koob, A.O.; Mante, M.; et al. TOM40 mediates mitochondrial dysfunction induced by alpha-synuclein accumulation in Parkinson's disease. *PLoS ONE* **2013**, *8*, e62277. [CrossRef]
63. Valente, E.M.; Abou-Sleiman, P.M.; Caputo, V.; Muqit, M.M.; Harvey, K.; Gispert, S.; Ali, Z.; Del Turco, D.; Bentivoglio, A.R.; Healy, D.G.; et al. Hereditary early-onset Parkinson's disease caused by mutations in PINK1. *Science* **2004**, *304*, 1158–1160. [CrossRef]
64. Pakpian, N.; Phopin, K.; Kitidee, K.; Govitrapong, P.; Wongchitrat, P. Alterations in Mitochondrial Dynamic-related Genes in the Peripheral Blood of Alzheimer's Disease Patients. *Curr. Alzheimer Res.* **2020**, *17*, 616–625. [CrossRef]
65. Carrocci, T.J.; Neugebauer, K.M. Pre-mRNA Splicing in the Nuclear Landscape. *Cold Spring Harb. Symp. Quant. Biol.* **2019**, *84*, 11–20. [CrossRef] [PubMed]
66. Agirre, E.; Oldfield, A.J.; Bellora, N.; Segelle, A.; Luco, R.F. Splicing-associated chromatin signatures: A combinatorial and position-dependent role for histone marks in splicing definition. *Nat. Commun.* **2021**, *12*, 682. [CrossRef] [PubMed]
67. Aslanzadeh, V.; Huang, Y.; Sanguinetti, G.; Beggs, J.D. Transcription rate strongly affects splicing fidelity and cotranscriptionality in budding yeast. *Genome Res.* **2018**, *28*, 203–213. [CrossRef]
68. Abebrese, E.L.; Ali, S.H.; Arnold, Z.R.; Andrews, V.M.; Armstrong, K.; Burns, L.; Crowder, H.R.; Day, R.T., Jr.; Hsu, D.G.; Jarrell, K.; et al. Identification of human short introns. *PLoS ONE* **2017**, *12*, e0175393. [CrossRef]
69. Kann, O.; Kovacs, R. Mitochondria and neuronal activity. *Am. J. Physiol. Cell Physiol.* **2007**, *292*, C641–C657. [CrossRef]
70. Chan, D.C. Mitochondria: Dynamic organelles in disease, aging, and development. *Cell* **2006**, *125*, 1241–1252. [CrossRef]
71. Castellani, C.A.; Longchamps, R.J.; Sun, J.; Guallar, E.; Arking, D.E. Thinking outside the nucleus: Mitochondrial DNA copy number in health and disease. *Mitochondrion* **2020**, *53*, 214–223. [CrossRef] [PubMed]
72. Yang, S.Y.; Castellani, C.A.; Longchamps, R.J.; Pillamarri, V.K.; O'Rourke, B.; Guallar, E.; Arking, D.E. Blood-derived mitochondrial DNA copy number is associated with gene expression across multiple tissues and is predictive for incident neurodegenerative disease. *Genome Res.* **2021**, *31*, 349–358. [CrossRef] [PubMed]
73. Longchamps, R.J.; Castellani, C.A.; Yang, S.Y.; Newcomb, C.E.; Sumpter, J.A.; Lane, J.; Grove, M.L.; Guallar, E.; Pankratz, N.; Taylor, K.D.; et al. Evaluation of mitochondrial DNA copy number estimation techniques. *PLoS ONE* **2020**, *15*, e0228166. [CrossRef]
74. Filograna, R.; Mennuni, M.; Alsina, D.; Larsson, N.G. Mitochondrial DNA copy number in human disease: The more the better? *FEBS Lett.* **2020**. [CrossRef]
75. Coskun, P.E.; Beal, M.F.; Wallace, D.C. Alzheimer's brains harbor somatic mtDNA control-region mutations that suppress mitochondrial transcription and replication. *Proc. Natl. Acad. Sci. USA* **2004**, *101*, 10726–10731. [CrossRef]
76. Rodriguez-Santiago, B.; Casademont, J.; Nunes, V. Is mitochondrial DNA depletion involved in Alzheimer's disease? *Eur. J. Hum. Genet.* **2001**, *9*, 279–285. [CrossRef]
77. Zamzami, N.; Marchetti, P.; Castedo, M.; Decaudin, D.; Macho, A.; Hirsch, T.; Susin, S.A.; Petit, P.X.; Mignotte, B.; Kroemer, G. Sequential reduction of mitochondrial transmembrane potential and generation of reactive oxygen species in early programmed cell death. *J. Exp. Med.* **1995**, *182*, 367–377. [CrossRef] [PubMed]
78. Yaniv, Y.; Juhaszova, M.; Nuss, H.B.; Wang, S.; Zorov, D.B.; Lakatta, E.G.; Sollott, S.J. Matching ATP supply and demand in mammalian heart: In vivo, in vitro, and in silico perspectives. *Ann. N. Y. Acad. Sci* **2010**, *1188*, 133–142. [CrossRef]
79. Zorov, D.B.; Juhaszova, M.; Sollott, S.J. Mitochondrial reactive oxygen species (ROS) and ROS-induced ROS release. *Physiol. Rev.* **2014**, *94*, 909–950. [CrossRef]
80. Izyumov, D.S.; Avetisyan, A.V.; Pletjushkina, O.Y.; Sakharov, D.V.; Wirtz, K.W.; Chernyak, B.V.; Skulachev, V.P. "Wages of fear": Transient threefold decrease in intracellular ATP level imposes apoptosis. *Biochim. Biophys. Acta* **2004**, *1658*, 141–147. [CrossRef] [PubMed]

Article

Shared Genetic Etiology between Alzheimer's Disease and Blood Levels of Specific Cytokines and Growth Factors

Robert J. van der Linden, Ward De Witte and Geert Poelmans *

Department of Human Genetics, Radboud University Medical Center, P.O. Box 9101, 6500 HB Nijmegen, The Netherlands; Robert.vanderLinden@radboudumc.nl (R.J.v.d.L.); Ward.deWitte@radboudumc.nl (W.D.W.)
* Correspondence: geert.poelmans@radboudumc.nl; Tel.: +31-630345956

Abstract: Late-onset Alzheimer's disease (AD) has a significant genetic and immunological component, but the molecular mechanisms through which genetic and immunity-related risk factors and their interplay contribute to AD pathogenesis are unclear. Therefore, we screened for genetic sharing between AD and the blood levels of a set of cytokines and growth factors to elucidate how the polygenic architecture of AD affects immune marker profiles. For this, we retrieved summary statistics from Finnish genome-wide association studies of AD and 41 immune marker blood levels and assessed for shared genetic etiology, using a polygenic risk score-based approach. For the blood levels of 15 cytokines and growth factors, we identified genetic sharing with AD. We also found positive and negative genetic concordances—implying that genetic risk factors for AD are associated with higher and lower blood levels—for several immune markers and were able to relate some of these results to the literature. Our results imply that genetic risk factors for AD also affect specific immune marker levels, which may be leveraged to develop novel treatment strategies for AD.

Keywords: Alzheimer's disease; cytokines and growth factors; genome-wide association study (GWAS); Polygenic Risk Score (PRS)-based analysis

1. Introduction

Alzheimer's disease (AD) is a neurodegenerative disorder that causes patients to suffer from behavioral changes, a progressive decline in memory and cognitive function due to brain atrophy resulting from neuronal loss of function and death [1]. While AD is accountable for most dementia cases and affects millions of people worldwide, no disease-modifying therapies are currently available [2]. The neuropathological hallmarks of AD are extracellular plaques of amyloid-β (Aβ) and intracellular neurofibrillary tangles (NFT) composed of excessively phosphorylated tau [3]. The heritability of AD is estimated at 60–80% [4]. Dominant mutations in *APP*, *PSEN1* and *PSEN2* cause rare familial forms of AD characterized by an early onset (early-onset AD (EOAD) < 65 years) [4]. However, in the vast majority of cases, AD symptoms only manifest later in life (late-onset AD (LOAD) ≥ 65 years) and multiple genetic risk factors with small effect sizes contribute to LOAD development [4]. The strongest common genetic risk factor for LOAD is the ε4 allele of the apolipoprotein E (APOE) gene (*APOEε4*) [4]. Furthermore, environmental risk factors contribute to the multifactorial nature of AD. The mechanisms through which these risk factors affect biological pathways that ultimately result in AD are largely unknown. Elucidating the polygenic architecture of AD may therefore provide insights for the development of disease-modifying therapies.

Many of the LOAD candidate genes that were identified through genome-wide association studies (GWASs) are thought to play a role in regulating immunity, through their effect on microglial function [5–7]. Microglial cells are the resident macrophages of the brain and in addition to Aβ plaques and tau tangles, increased microglial activity and associated neuroinflammation has emerged as a third core pathology in AD [8]. Under physiological conditions, microglial cells survey the brain but only become activated and cause inflammation upon recognizing threats, such as infection, toxins and injury [8]. Although acute neuroinflammation could still serve as a defense mechanism against these threats, chronic neuroinflammation by excessively active microglial cells and recruitment of peripheral macrophages is detrimental to neuronal function [8]. The effects of peripheral macrophages may be exacerbated by impaired function of the blood–brain barrier (BBB) that separates the central nervous system from the rest of the body, and breakdown of the BBB is often observed in AD [9]. Moreover, activated microglial cells contribute to these detrimental effects by massively releasing inflammatory molecules and excessively pruning synapses, leading to synaptic loss [10]. In this respect, AD patient brains contain higher levels of activated, pro-inflammatory microglial cells [11,12]. In addition, the sustained activation of microglia and other immune cells has been demonstrated to exacerbate both $A\beta$ and tau pathology [5,8]. Furthermore, genetic pleiotropy analyses have identified genetic overlap between AD and immune-mediated diseases—i.e., Crohn's disease, psoriasis, and type 1 diabetes—indicating that aberrant immune processes influence AD pathogenesis and progression [12]. Chronic neuroinflammation in AD can be caused by both overexpression of pro-inflammatory cytokines and downregulation of anti-inflammatory cytokines that neutralize the harmful effects of chronic exposure to pro-inflammatory cytokines (reviewed in [13]). Moreover, the dysregulated immune system in AD is not limited to the central nervous system, and there is ample evidence for systemic immune signals (originating from outside the brain) contributing to AD (reviewed in [14]). Although all these data suggest a relationship between inflammation and AD, a (much) better understanding of this relationship could have implications for treatment and prevention strategies.

In this study, we investigated whether there is overlap between genetic risk factors—single nucleotide polymorphisms (SNPs)—for LOAD and SNPs contributing to the blood levels of a set of immune markers (cytokines and growth factors, inflammatory regulators that can be used as important intermediate phenotypes for inflammatory diseases [15]). To this end, we deployed shared genetic etiology and SNP effect concordance analyses, using publicly available GWAS results.

2. Materials and Methods
2.1. GWAS Summary Statistics for PRS-Based Analyses

For the polygenic risk score (PRS)-based analyses (see below), we used GWAS summary statistics from a Finnish AD cohort (1798 cases (diagnosed with ICD-10 code G301), 72,206 healthy controls) obtained through FinnGen (finngen_r4_AD_LO_EXMORE) as the 'base' sample. For the 'target' samples, we retrieved GWAS summary statistics for the blood levels of 41 immune markers (cytokines and growth factors) that were measured as a continuous phenotype in the general population from the study by Ahola-Olli et al. (GWAS sizes ranging from 840 to 8293 subjects; Table 1) [15].

Table 1. GWAS summary statistics for the blood levels of 41 immune markers in the general population that were used in this study were obtained from the study by Ahola-Olli et al.

Name	Description	Type	N GWAS
CCL11	Eotaxin	Cytokine	8153
CCL2 (MCP1)	Monocyte chemotactic protein-1	Cytokine	8293
CCL27 (CTACK)	Cutaneous T-cell attracting	Cytokine	3631
CCL3 (MIP1α)	Macrophage inflammatory protein-1α	Cytokine	3522
CCL4 (MIP1β)	Macrophage inflammatory protein-1β	Cytokine	8243
CCL5 (RANTES)	Regulated upon activation, normal T cell expressed and secreted	Cytokine	3421
CCL7 (MCP3)	Monocyte specific chemokine 3	Cytokine	843
CXCL1 (GROa)	Growth-regulated oncogene-α	Cytokine	3505
CXCL10 (IP10)	Interferon γ-induced protein 10	Cytokine	3685
CXCL12 (SDF1a)	Stromal cell-derived factor-1 α	Cytokine	5998
CXCL9 (MIG)	Monokine induced by interferon-γ	Cytokine	3685

Table 1. Cont.

Name	Description	Type	N GWAS
FGF2 (FGFBasic)	Basic fibroblast growth factor	Growth factor	7565
GCSF	Granulocyte colony-stimulating factor	Growth factor	7904
HGF	Hepatocyte growth factor	Growth factor	8292
IFNγ	Interferon-γ	Cytokine	7701
IL10	Interleukin-10	Cytokine	7681
IL12p70	Interleukin-12p70	Cytokine	8270
IL13	Interleukin-13	Cytokine	3557
IL16	Interleukin-16	Cytokine	3483
IL17	Interleukin-17	Cytokine	7760
IL18	Interleukin-18	Cytokine	3636
IL1b	Interleukin-1-β	Cytokine	3309
IL1ra	Interleukin-1 receptor antagonist	Cytokine	3638
IL2	Interleukin-2	Cytokine	3475
IL2ra	Interleukin-2 receptor, α subunit	Cytokine	3677
IL4	Interleukin-4	Cytokine	8124
IL5	Interleukin-5	Cytokine	3364
IL6	Interleukin-6	Cytokine	8189
IL7	Interleukin-7	Cytokine	3409
IL8 (CXCL8)	Interleukin-8	Cytokine	3526
IL9	Interleukin-9	Cytokine	3634
MCSF	Macrophage colony-stimulating factor	Growth factor	840
MIF	Macrophage migration inhibitory factor (glycosylation-inhibiting factor)	Growth factor	3494
PDGFbb	Platelet-derived growth factor BB	Growth factor	8293
SCF	Stem cell factor	Growth factor	8290
SCGFβ	Stem cell growth factor β	Growth factor	3682
TNFα	Tumor necrosis factor-α	Growth factor	3454
TNFβ	Tumor necrosis factor-β	Growth factor	1559
TRAIL	TNF-related apoptosis-inducing ligand	Cytokine	8186
VEGF	Vascular endothelial growth factor	Growth factor	7118
βNGF	β nerve growth factor	Growth factor	3531

NOTE: Alternative names are indicated between brackets. Abbreviation: GWAS, genome-wide association study.

2.2. PRS-Based Analyses

We performed PRS-based analyses with PRSice (v1.25), using the abovementioned Finish GWAS summary statistics as 'base' and 'target' samples [16] to determine the level of shared genetic etiology between AD and the blood levels of the 41 immune markers. First, we performed clumping based on the p-values of SNPs in the 'base sample' to select the most significant SNP among correlated SNPs that were in linkage disequilibrium (LD, $R^2 > 0.25$) within a window of 500 kb using PLINK (v1.90) [17,18]. With PRSice we then calculated summary-level PRS by regressing the weights of selected AD risk SNPs (based on their p-value in the AD GWAS) on to the calculated weighted multi-SNP risk scores of immune marker blood levels, using the gtx package implemented in PRSice [16]. The PRS-based analyses were performed for all SNPs that exceeded seven default p-value thresholds (P_T): 0.001, 0.05, 0.1, 0.2, 0.3, 0.4 and 0.5. A correction for multiple testing was then performed using a Bonferroni significance threshold < 0.05 (i.e., $p < 0.05/287$ tests (=7 P_Ts × 41 phenotypes) = 1.74×10^{-4}).

2.3. SNP Effect Concordance Analyses

Subsequently, for those immune markers that we found to have a shared genetic etiology with AD, we performed SNP effect concordance analysis (SECA) to determine the direction of the genetic overlap [19]. We used SECA to calculate empirical p-values for the concordance between AD and immune marker blood levels, i.e., the agreement in the direction of the SNP effect across two phenotypes. We then performed a Bonferroni correction to account for the number of tests that we performed with SECA.

3. Results

3.1. PRS-Based Analyses

After correcting for multiple testing, we identified genetic sharing between AD and the blood levels of 15 out of the 41 immune markers (36.6%) for at least one of the seven used p-value thresholds (P_Ts) (Table 2). A complete overview of the PRS-based analyses including all P_Ts for all blood immune markers is provided in the Supplementary Materials (Table S1). For the 15 significant immune markers, genetic variants associated with AD also explained between 0.5 and 1.8% of the variation in their blood levels (Table 2).

Table 2. Fifteen immune markers for which we identified genetic sharing between AD and their blood levels.

Immune Marker	Best P_T	N SNPs	Bonferroni p-Value	Variance Explained R^2	Concordance with AD
CCL4 (MIP1β)	0.001	2001	3.51×10^{-5}	0.003227	+
FGF2 (FGFBasic)	0.5	497,296	7.18×10^{-7}	0.004509	+
GCSF	0.2	236,031	3.45×10^{-3}	0.002255	+
HGF	0.5	498,823	4.25×10^{-4}	0.002631	+
IL10	0.5	497,485	1.38×10^{-7}	0.004857	+
IL12p70	0.5	498,327	7.70×10^{-3}	0.001971	+
SCF	0.5	498,841	3.14×10^{-3}	0.002171	+
bNGF	0.001	1894	4.03×10^{-3}	0.004956	-
CCL3 (MIP1α)	0.05	65,371	2.25×10^{-2}	0.004048	-
CCL5 (RANTES)	0.3	310,864	2.45×10^{-8}	0.011848	-
CXCL1 (GROα)	0.2	221,410	1.54×10^{-13}	0.018177	-
IL8	0.2	220,931	9.84×10^{-5}	0.006968	-
MIF	0.2	221,727	6.80×10^{-5}	0.007234	-
SCGFβ	0.001	1927	1.54×10^{-4}	0.006441	-
TRAIL	0.3	332,200	3.18×10^{-5}	0.003273	-

Abbreviations: P_T, p-value threshold; SNP, single nucleotide polymorphism.

3.2. SECA Analyses

SECA analyses showed a significant genetic concordance between AD and all 15 immune markers that emerged from our screening (Table S2). For the blood levels of seven of the 15 immune markers, we identified a positive concordance with AD, indicating that genetic risk factors associated with AD also contribute to increased blood levels of these markers (Table 2). For the other eight immune markers, we found a negative concordance with AD, implying that genetic risk factors for AD are also associated with lower blood levels of these markers (Table 2).

4. Discussion

In this paper, we identified genetic sharing between AD and the blood levels of 15 immune markers. Through concordance analyses, we also determined that for eight and seven of these markers, AD genetic risk factors contribute to increased and decreased blood levels, respectively.

First, we will discuss the literature findings about those immune markers for which we found a positive concordance between AD and their blood levels. Although no direct links between the blood levels of basic fibroblast growth factor (bFGF; other name: FGF2) have been reported, increased FGF2 levels were found in the brains of AD patients. In these brains, FGF2 was found within the neuritic plaques and in association with the neurofibrillary tangles that are characteristic of AD [20]. Further, FGF2 gene transfer restores hippocampal functions in mouse models of AD and viral delivery of FGF2 in the brain has been proposed as a therapeutic intervention for AD [21], further indicating an important role for FGF2 in AD. Granulocyte colony-stimulating factor (GCSF) stimulates the production and release of neutrophils in the blood and is also a neurotrophic factor [22]. In this respect, it is interesting that decreased GCSF blood levels have been reported in

AD patients, although among these patients, higher GSCF blood levels associate with increased disease severity [22]. However, while this latter finding is in keeping with the positive genetic concordance between AD and blood GCSF levels that we observed, GCSF treatment has also been shown to improve memory in an AD rat model [23]. In addition, again in line with our findings, higher levels of hepatocyte growth factor (HGF)—which regulates various brain functions, including axonal outgrowth, neuronal survival, and synaptic plasticity—have been found in the blood, cerebrospinal fluid (CSF) and brains of AD patients [24,25]. In this respect, increased HGF immunoreactivity within neurons, astrocytes and microglial cells was also demonstrated to be an indicator of gliosis and microglial proliferation that occurs around Aβ plaques in AD brains [25]. In contrast to the results of our concordance analysis, decreased blood levels of stem cell factor (SCF) have been described in AD patients, and these decreased levels are also associated with a higher rate of cognitive decline [26].

In addition to the four abovementioned growth factors, we identified a positive genetic concordance between AD and the blood levels of three cytokines: interleukin 10 (IL10), IL12p70 and CCL4. As for IL10, this cytokine is a negative regulator of the innate immune system and *IL10* knockout in an AD mouse model resulted in increased Aβ clearance by activated microglia and a partially rescued synaptic integrity in the brains of these mice [27]. The negative role of IL10 in AD is corroborated by our finding that genetic variants associated with AD also contribute to increased IL10 blood levels. Furthermore, in the brains of AD patients, significantly increased levels of both the anti-inflammatory IL10 and the pro-inflammatory IL12p70 have been reported, indicating that both anti- and pro-inflammatory signaling can be activated simultaneously in AD [28]. In addition, IL12p70 has been shown to reduce neuronal viability in cell culture experiments, both in the presence or absence of Aβ [29]. Lastly, chronic inflammation leads to elevated CCL4 levels in AD brains [30], while CCL4 levels are also increased in microglia associated with Aβ plaques [31].

For eight immune markers (three growth factors and five cytokines), we found a negative concordance between AD and their blood levels. Firstly, nerve growth factor (NGF) contributes to the survival, regeneration, and death of neurons during aging and in neurodegenerative diseases such as AD [32]. Impaired NGF signaling has also been linked to neurons losing their cholinergic phenotype in the AD basal forebrain [33] and brain implants delivering NGF to the cholinergic basal forebrain are currently being tested as an AD treatment in humans [34]. In this respect, it is interesting that we found that genetic variants associated with AD contribute to decreased NGF levels in the blood, which may reflect what is happening with NGF levels in the brain. Further, macrophage migration inhibitory factor (MIF) levels have been found to be increased in the blood, CSF and brains of AD patients [35]. Although the finding about MIF levels in the blood of AD patients is opposite to the negative genetic concordance that we identified, MIF colocalizes with Aβ plaques and increased MIF levels protect neuronal cells from Aβ-induced neurotoxicity [36]. Hence, MIF levels may be upregulated in the brain as a defense mechanism to compensate for declined cognitive function in AD [36]. Moreover, although no direct links between blood SCGFβ levels and AD have been reported, it is interesting that blood IL10 levels—for which we found a positive concordance with AD—negatively predict blood SCGFβ levels [37], which is in line with the negative concordance with AD that we observed.

As for the cytokines for which we found a negative concordance between AD and their blood levels, microvessels from AD brains produce and release high levels of CCL3 compared to control brains, suggesting that the brain microvasculature contributes to the inflammatory environment of the AD brain through upregulating CCL3 expression [38]. In addition, elevated levels of CCL5 in the cerebral microcirculation of AD patients were reported, and CCL5 treatment of neurons increases cell survival, suggesting a neuroprotective role for CCL5 [39,40]. Moreover, monocytes from AD patients produce significantly higher amounts of CXCL1 compared to age-matched controls, which causes these monocytes to migrate from the blood to the brain [41]. In the brain, CXCL1 also promotes the cleavage of tau, which is considered an early event in AD development [42]. Taken together, the literature findings on the relationship between AD and the brain levels of CCL3, CCL5 and CXCL1 are opposed to the negative concordance between AD and their blood levels that we found. In this respect, there may be an inverse relationship between the blood and brain levels of these cytokines, which we could speculate would, e.g., result from the recruitment of cytokine-producing immune cells to the site of inflammation—i.e., from the blood to the brain in the case of AD—that would in turn lead to a relative depletion of the cytokines in the blood. Incidentally, this may also apply to the other immune markers for which we found a discrepancy between the literature findings and the results from our concordance analysis—such as SCF, see above—and in fact, a recent study found that there were relatively few direct correlations between blood and CSF levels of cytokines in multiple neuro-inflammatory diseases [43]. Further, TRAIL is specifically expressed in the brains of AD patients and completely absent in the brains of healthy controls [44], and anti-TRAIL antibodies reduce brain Aβ load and improve cognition in an AD mouse model [45]. However, it was also reported that TRAIL blood levels do not differ between AD patients and controls [46]. Lastly, blood IL8 levels were found to be decreased in AD patients—consistent with the negative genetic concordance that we observed—while both lower and higher IL8 levels have been reported in the CSF of AD patients [47,48]. In addition, IL8 promotes inflammation and cell death of cultured neurons [49] while, in contrast, neurons also produce IL8 as a protection against Aβ-induced toxicity [50].

This study has two main limitations. First, genetic sharing between AD and blood levels of immune markers does not necessarily mean higher or lower immune marker blood levels are causative of AD. Second, as already indicated, blood levels may not reflect what is happening in the brain directly and there may also be an inverse relationship between the levels of immune markers in the blood and CSF, which warrants further investigation using CSF and post-mortem brain samples. This being said, we can conclude that genetic risk factors for AD also affect the blood levels of specific immune markers, suggesting that systemic immune processes may influence AD pathogenesis and progression. Although further studies are needed to confirm our findings, and depending on whether AD shows genetic overlap with increased or decreased immune marker blood levels, novel treatment strategies for AD could be developed.

Supplementary Materials: The following are available online at https://www.mdpi.com/article/10.3390/genes12060865/s1, Table S1: Shared genetic etiology analyses between AD and the blood levels of 41 immune markers across seven p-value inclusion thresholds (PTs) using a polygenic risk score (PRS)-based approach, Table S2: SNP effect concordance (SECA) analyses on the genetic concordance between AD and blood immune marker levels that were significant in the polygenic risk score (PRS)-based analyses.

Author Contributions: R.J.v.d.L. and G.P.: Conceptualization; R.J.v.d.L. and W.D.W.: Investigation, Formal Analysis, Data Curation; R.J.v.d.L.: Writing—Original Draft; G.P.: Writing—Review and Editing; G.P.: Supervision; G.P.: Funding Acquisition. All authors have read and agreed to the published version of the manuscript.

Funding: The research presented in this paper was supported by the Dutch charity foundation 'Stichting Devon'.

Institutional Review Board Statement: Not applicable.

Informed Consent Statement: Not applicable.

Data Availability Statement: FinnGen Alzheimer's disease GWAS summary statistics can be accessed online at: https://www.finngen.fi/en/access_results (accessed on 11 May 2021). Cytokine and growth factor GWAS summary statistics from the study by Ahola-Olli et al. can be accessed online at: https://www.ebi.ac.uk/gwas/publications/27989323 (accessed on 11 May 2021).

Acknowledgments: We want to acknowledge the participants and investigators of the FinnGen study.

Conflicts of Interest: G.P. is the director of Drug Target ID., Ltd. (Nijmegen, The Netherlands). The other authors have no competing interests to declare.

References

1. Livingston, G.; Huntley, J.; Sommerlad, A.; Ames, D.; Ballard, C.; Banerjee, S.; Brayne, C.; Burns, A.; Cohen-Mansfield, J.; Cooper, C.; et al. Dementia prevention, intervention, and care: 2020 report of the Lancet Commission. *Lancet* **2020**, *396*, 413–446. [CrossRef]
2. Patterson, C. *World Alzheimer Report 2018: The State of the Art of Dementia Research: New Frontiers*; Alzheimer's Disease International (ADI): London, UK, 2018.
3. Jack, C.R., Jr.; Bennett, D.A.; Blennow, K.; Carrillo, M.C.; Dunn, B.; Haeberlein, S.B.; Holtzman, D.M.; Jagust, W.; Jessen, F.; Karlawish, J.; et al. NIA-AA Research Framework: Toward a biological definition of Alzheimer's disease. *Alzheimers Dement.* **2018**, *14*, 535–562. [CrossRef]
4. Bekris, L.M.; Yu, C.E.; Bird, T.D.; Tsuang, D.W. Genetics of Alzheimer disease. *J. Geriatr. Psychiatry Neurol.* **2010**, *23*, 213–227. [CrossRef] [PubMed]
5. Heneka, M.T.; Carson, M.J.; El Khoury, J.; Landreth, G.E.; Brosseron, F.; Feinstein, D.L.; Jacobs, A.H.; Wyss-Coray, T.; Vitorica, J.; Ransohoff, R.M.; et al. Neuroinflammation in Alzheimer's disease. *Lancet Neurol.* **2015**, *14*, 388–405. [CrossRef]
6. Jansen, I.E.; Savage, J.E.; Watanabe, K.; Bryois, J.; Williams, D.M.; Steinberg, S.; Sealock, J.; Karlsson, I.K.; Hagg, S.; Athanasiu, L.; et al. Genome-wide meta-analysis identifies new loci and functional pathways influencing Alzheimer's disease risk. *Nat. Genet.* **2019**, *51*, 404–413. [CrossRef]
7. Kunkle, B.W.; Grenier-Boley, B.; Sims, R.; Bis, J.C.; Damotte, V.; Naj, A.C.; Boland, A.; Vronskaya, M.; van der Lee, S.J.; Amlie-Wolf, A.; et al. Genetic meta-analysis of diagnosed Alzheimer's disease identifies new risk loci and implicates Abeta, tau, immunity and lipid processing. *Nat. Genet.* **2019**, *51*, 414–430. [CrossRef]
8. Kinney, J.W.; Bemiller, S.M.; Murtishaw, A.S.; Leisgang, A.M.; Salazar, A.M.; Lamb, B.T. Inflammation as a central mechanism in Alzheimer's disease. *Alzheimers Dement. (N. Y.)* **2018**, *4*, 575–590. [CrossRef] [PubMed]
9. Sweeney, M.D.; Sagare, A.P.; Zlokovic, B.V. Blood-brain barrier breakdown in Alzheimer disease and other neurodegenerative disorders. *Nat. Rev. Neurol.* **2018**, *14*, 133–150. [CrossRef]
10. Salter, M.W.; Stevens, B. Microglia emerge as central players in brain disease. *Nat. Med.* **2017**, *23*, 1018–1027. [CrossRef] [PubMed]
11. Remarque, E.J.; Bollen, E.L.; Weverling-Rijnsburger, A.W.; Laterveer, J.C.; Blauw, G.J.; Westendorp, R.G. Patients with Alzheimer's disease display a pro-inflammatory phenotype. *Exp. Gerontol.* **2001**, *36*, 171–176. [CrossRef]
12. Yokoyama, J.S.; Wang, Y.; Schork, A.J.; Thompson, W.K.; Karch, C.M.; Cruchaga, C.; McEvoy, L.K.; Witoelar, A.; Chen, C.H.; Holland, D.; et al. Association Between Genetic Traits for Immune-Mediated Diseases and Alzheimer Disease. *JAMA Neurol.* **2016**, *73*, 691–697. [CrossRef]
13. Su, F.; Bai, F.; Zhang, Z. Inflammatory Cytokines and Alzheimer's Disease: A Review from the Perspective of Genetic Polymorphisms. *Neurosci. Bull.* **2016**, *32*, 469–480. [CrossRef] [PubMed]
14. Cao, W.; Zheng, H. Peripheral immune system in aging and Alzheimer's disease. *Mol. Neurodegener.* **2018**, *13*, 51. [CrossRef] [PubMed]
15. Ahola-Olli, A.V.; Wurtz, P.; Havulinna, A.S.; Aalto, K.; Pitkanen, N.; Lehtimaki, T.; Kahonen, M.; Lyytikainen, L.P.; Raitoharju, E.; Seppala, I.; et al. Genome-wide Association Study Identifies 27 Loci Influencing Concentrations of Circulating Cytokines and Growth Factors. *Am. J. Hum. Genet.* **2017**, *100*, 40–50. [CrossRef] [PubMed]
16. Euesden, J.; Lewis, C.M.; O'Reilly, P.F. PRSice: Polygenic Risk Score software. *Bioinformatics* **2015**, *31*, 1466–1468. [CrossRef]
17. Bralten, J.; van Hulzen, K.J.; Martens, M.B.; Galesloot, T.E.; Arias Vasquez, A.; Kiemeney, L.A.; Buitelaar, J.K.; Muntjewerff, J.W.; Franke, B.; Poelmans, G. Autism spectrum disorders and autistic traits share genetics and biology. *Mol. Psychiatry* **2018**, *23*, 1205–1212. [CrossRef] [PubMed]
18. Xicoy, H.; Klemann, C.J.; De Witte, W.; Martens, M.B.; Martens, G.J.; Poelmans, G. Shared genetic etiology between Parkinson's disease and blood levels of specific lipids. *NPJ Parkinsons Dis.* **2021**, *7*, 23. [CrossRef]
19. Nyholt, D.R. SECA: SNP effect concordance analysis using genome-wide association summary results. *Bioinformatics* **2014**, *30*, 2086–2088. [CrossRef] [PubMed]
20. Stopa, E.G.; Gonzalez, A.M.; Chorsky, R.; Corona, R.J.; Alvarez, J.; Bird, E.D.; Baird, A. Basic fibroblast growth factor in Alzheimer's disease. *Biochem. Biophys. Res. Commun.* **1990**, *171*, 690–696. [CrossRef]
21. Kiyota, T.; Ingraham, K.L.; Jacobsen, M.T.; Xiong, H.; Ikezu, T. FGF2 gene transfer restores hippocampal functions in mouse models of Alzheimer's disease and has therapeutic implications for neurocognitive disorders. *Proc. Natl. Acad. Sci. USA* **2011**, *108*, E1339–E1348. [CrossRef]
22. Barber, R.C.; Edwards, M.I.; Xiao, G.; Huebinger, R.M.; Diaz-Arrastia, R.; Wilhelmsen, K.C.; Hall, J.R.; O'Bryant, S.E. Serum granulocyte colony-stimulating factor and Alzheimer's disease. *Dement. Geriatr. Cogn. Dis. Extra* **2012**, *2*, 353–360. [CrossRef]
23. Prakash, A.; Medhi, B.; Chopra, K. Granulocyte colony stimulating factor (GCSF) improves memory and neurobehavior in an amyloid-beta induced experimental model of Alzheimer's disease. *Pharmacol. Biochem. Behav.* **2013**, *110*, 46–57. [CrossRef] [PubMed]

24. Zhu, Y.; Hilal, S.; Chai, Y.L.; Ikram, M.K.; Venketasubramanian, N.; Chen, C.P.; Lai, M.K.P. Serum Hepatocyte Growth Factor Is Associated with Small Vessel Disease in Alzheimer's Dementia. *Front. Aging Neurosci.* **2018**, *10*, 8. [CrossRef] [PubMed]
25. Fenton, H.; Finch, P.W.; Rubin, J.S.; Rosenberg, J.M.; Taylor, W.G.; Kuo-Leblanc, V.; Rodriguez-Wolf, M.; Baird, A.; Schipper, H.M.; Stopa, E.G. Hepatocyte growth factor (HGF/SF) in Alzheimer's disease. *Brain Res.* **1998**, *779*, 262–270. [CrossRef]
26. Laske, C.; Sopova, K.; Hoffmann, N.; Stransky, E.; Hagen, K.; Fallgatter, A.J.; Stellos, K.; Leyhe, T. Stem cell factor plasma levels are decreased in Alzheimer's disease patients with fast cognitive decline after one-year follow-up period: The Pythia-study. *J. Alzheimers Dis.* **2011**, *26*, 39–45. [CrossRef]
27. Guillot-Sestier, M.V.; Doty, K.R.; Gate, D.; Rodriguez, J., Jr.; Leung, B.P.; Rezai-Zadeh, K.; Town, T. Il10 deficiency rebalances innate immunity to mitigate Alzheimer-like pathology. *Neuron* **2015**, *85*, 534–548. [CrossRef]
28. Franco-Bocanegra, D.K.; George, B.; Lau, L.C.; Holmes, C.; Nicoll, J.A.R.; Boche, D. Microglial motility in Alzheimer's disease and after Abeta42 immunotherapy: A human post-mortem study. *Acta Neuropathol. Commun.* **2019**, *7*, 174. [CrossRef]
29. Wood, L.B.; Winslow, A.R.; Proctor, E.A.; McGuone, D.; Mordes, D.A.; Frosch, M.P.; Hyman, B.T.; Lauffenburger, D.A.; Haigis, K.M. Identification of neurotoxic cytokines by profiling Alzheimer's disease tissues and neuron culture viability screening. *Sci. Rep.* **2015**, *5*, 16622. [CrossRef] [PubMed]
30. Zhu, M.; Allard, J.S.; Zhang, Y.; Perez, E.; Spangler, E.L.; Becker, K.G.; Rapp, P.R. Age-related brain expression and regulation of the chemokine CCL4/MIP-1beta in APP/PS1 double-transgenic mice. *J. Neuropathol. Exp. Neurol.* **2014**, *73*, 362–374. [CrossRef]
31. Yin, Z.; Raj, D.; Saiepour, N.; Van Dam, D.; Brouwer, N.; Holtman, I.R.; Eggen, B.J.L.; Moller, T.; Tamm, J.A.; Abdourahman, A.; et al. Immune hyperreactivity of Abeta plaque-associated microglia in Alzheimer's disease. *Neurobiol. Aging* **2017**, *55*, 115–122. [CrossRef]
32. Xu, C.J.; Wang, J.L.; Jin, W.L. The Emerging Therapeutic Role of NGF in Alzheimer's Disease. *Neurochem. Res.* **2016**, *41*, 1211–1218. [CrossRef] [PubMed]
33. Cuello, A.C.; Pentz, R.; Hall, H. The Brain NGF Metabolic Pathway in Health and in Alzheimer's Pathology. *Front. Neurosci.* **2019**, *13*, 62. [CrossRef]
34. Eyjolfsdottir, H.; Eriksdotter, M.; Linderoth, B.; Lind, G.; Juliusson, B.; Kusk, P.; Almkvist, O.; Andreasen, N.; Blennow, K.; Ferreira, D.; et al. Targeted delivery of nerve growth factor to the cholinergic basal forebrain of Alzheimer's disease patients: Application of a second-generation encapsulated cell biodelivery device. *Alzheimers Res. Ther.* **2016**, *8*, 30. [CrossRef]
35. Azizi, G.; Khannazer, N.; Mirshafiey, A. The Potential Role of Chemokines in Alzheimer's Disease Pathogenesis. *Am. J. Alzheimers Dis. Other Demen.* **2014**, *29*, 415–425. [CrossRef] [PubMed]
36. Zhang, S.; Zhao, J.; Zhang, Y.; Zhang, Y.; Cai, F.; Wang, L.; Song, W. Upregulation of MIF as a defense mechanism and a biomarker of Alzheimer's disease. *Alzheimers Res. Ther.* **2019**, *11*, 54. [CrossRef]
37. Tarantino, G.; Citro, V.; Balsano, C.; Capone, D. Could SCGF-Beta Levels Be Associated with Inflammation Markers and Insulin Resistance in Male Patients Suffering from Obesity-Related NAFLD? *Diagnostics (Basel)* **2020**, *10*, 395. [CrossRef]
38. Tripathy, D.; Thirumangalakudi, L.; Grammas, P. Expression of macrophage inflammatory protein 1-alpha is elevated in Alzheimer's vessels and is regulated by oxidative stress. *J. Alzheimers Dis.* **2007**, *11*, 447–455. [CrossRef]
39. Vacinova, G.; Vejrazkova, D.; Rusina, R.; Holmerova, I.; Vankova, H.; Jarolimova, E.; Vcelak, J.; Bendlova, B.; Vankova, M. Regulated upon activation, normal T cell expressed and secreted (RANTES) levels in the peripheral blood of patients with Alzheimer's disease. *Neural Regen. Res.* **2021**, *16*, 796–800. [CrossRef]
40. Tripathy, D.; Thirumangalakudi, L.; Grammas, P. RANTES upregulation in the Alzheimer's disease brain: A possible neuroprotective role. *Neurobiol. Aging* **2010**, *31*, 8–16. [CrossRef] [PubMed]
41. Zhang, K.; Tian, L.; Liu, L.; Feng, Y.; Dong, Y.B.; Li, B.; Shang, D.S.; Fang, W.G.; Cao, Y.P.; Chen, Y.H. CXCL1 contributes to beta-amyloid-induced transendothelial migration of monocytes in Alzheimer's disease. *PLoS ONE* **2013**, *8*, e72744. [CrossRef]
42. Zhang, X.F.; Zhao, Y.F.; Zhu, S.W.; Huang, W.J.; Luo, Y.; Chen, Q.Y.; Ge, L.J.; Li, R.S.; Wang, J.F.; Sun, M.; et al. CXCL1 Triggers Caspase-3 Dependent Tau Cleavage in Long-Term Neuronal Cultures and in the Hippocampus of Aged Mice: Implications in Alzheimer's Disease. *J. Alzheimers Dis.* **2015**, *48*, 89–104. [CrossRef] [PubMed]
43. Lepennetier, G.; Hracsko, Z.; Unger, M.; Van Griensven, M.; Grummel, V.; Krumbholz, M.; Berthele, A.; Hemmer, B.; Kowarik, M.C. Cytokine and immune cell profiling in the cerebrospinal fluid of patients with neuro-inflammatory diseases. *J. Neuroinflammation* **2019**, *16*, 219. [CrossRef] [PubMed]
44. Uberti, D.; Cantarella, G.; Facchetti, F.; Cafici, A.; Grasso, G.; Bernardini, R.; Memo, M. TRAIL is expressed in the brain cells of Alzheimer's disease patients. *Neuroreport* **2004**, *15*, 579–581. [CrossRef]
45. Cantarella, G.; Di Benedetto, G.; Puzzo, D.; Privitera, L.; Loreto, C.; Saccone, S.; Giunta, S.; Palmeri, A.; Bernardini, R. Neutralization of TNFSF10 ameliorates functional outcome in a murine model of Alzheimer's disease. *Brain* **2015**, *138*, 203–216. [CrossRef]
46. Genc, S.; Egrilmez, M.Y.; Yaka, E.; Cavdar, Z.; Iyilikci, L.; Yener, G.; Genc, K. TNF-related apoptosis-inducing ligand level in Alzheimer's disease. *Neurol. Sci.* **2009**, *30*, 263–267. [CrossRef]
47. Hesse, R.; Wahler, A.; Gummert, P.; Kirschmer, S.; Otto, M.; Tumani, H.; Lewerenz, J.; Schnack, C.; von Arnim, C.A. Decreased IL-8 levels in CSF and serum of AD patients and negative correlation of MMSE and IL-1beta. *BMC Neurol.* **2016**, *16*, 185. [CrossRef]
48. Galimberti, D.; Schoonenboom, N.; Scheltens, P.; Fenoglio, C.; Bouwman, F.; Venturelli, E.; Guidi, I.; Blankenstein, M.A.; Bresolin, N.; Scarpini, E. Intrathecal chemokine synthesis in mild cognitive impairment and Alzheimer disease. *Arch. Neurol.* **2006**, *63*, 538–543. [CrossRef]

49. Thirumangalakudi, L.; Yin, L.; Rao, H.V.; Grammas, P. IL-8 induces expression of matrix metalloproteinases, cell cycle and pro-apoptotic proteins, and cell death in cultured neurons. *J. Alzheimers Dis.* **2007**, *11*, 305–311. [CrossRef] [PubMed]
50. Ashutosh; Kou, W.; Cotter, R.; Borgmann, K.; Wu, L.; Persidsky, R.; Sakhuja, N.; Ghorpade, A. CXCL8 protects human neurons from amyloid-beta-induced neurotoxicity: Relevance to Alzheimer's disease. *Biochem. Biophys. Res. Commun.* **2011**, *412*, 565–571. [CrossRef]

Article

The Impact of Complement Genes on the Risk of Late-Onset Alzheimer's Disease

Sarah M. Carpanini [1,2,†], Janet C. Harwood [3,†], Emily Baker [1], Megan Torvell [1,2], The GERAD1 Consortium [‡], Rebecca Sims [3], Julie Williams [1] and B. Paul Morgan [1,2,*]

1 UK Dementia Research Institute at Cardiff University, School of Medicine, Cardiff, CF24 4HQ, UK; CarpaniniS@Cardiff.ac.uk (S.M.C.); BakerEA@cardiff.ac.uk (E.B.); TorvellM@cardiff.ac.uk (M.T.); WilliamsJ@cardiff.ac.uk (J.W.)
2 Division of Infection and Immunity, School of Medicine, Systems Immunity Research Institute, Cardiff University, Cardiff, CF14 4XN, UK
3 Division of Psychological Medicine and Clinical Neurosciences, School of Medicine, Cardiff University, Cardiff, CF24 4HQ, UK; HarwoodJC@cardiff.ac.uk (J.C.H.); SimsRC@cardiff.ac.uk (R.S.)
* Correspondence: Morganbp@cardiff.ac.uk
† These authors contributed equally to this work.
‡ Data used in the preparation of this article were obtained from the Genetic and Environmental Risk for Alzheimer's disease (GERAD1) Consortium. As such, the investigators within the GERAD1 consortia contributed to the design and implementation of GERAD1 and/or provided data but did not participate in analysis or writing of this report. A full list of GERAD1 investigators and their affiliations is included in Supplementary File S1.

Citation: Carpanini, S.M.; Harwood, J.C.; Baker, E.; Torvell, M.; The GERAD1 Consortium; Sims, R.; Williams, J.; Morgan, B.P. The Impact of Complement Genes on the Risk of Late-Onset Alzheimer's Disease. *Genes* **2021**, *12*, 443. https://doi.org/10.3390/genes12030443

Academic Editors: Laura Ibanez and Justin Miller

Received: 27 February 2021
Accepted: 16 March 2021
Published: 20 March 2021

Publisher's Note: MDPI stays neutral with regard to jurisdictional claims in published maps and institutional affiliations.

Copyright: © 2021 by the authors. Licensee MDPI, Basel, Switzerland. This article is an open access article distributed under the terms and conditions of the Creative Commons Attribution (CC BY) license (https://creativecommons.org/licenses/by/4.0/).

Abstract: Late-onset Alzheimer's disease (LOAD), the most common cause of dementia, and a huge global health challenge, is a neurodegenerative disease of uncertain aetiology. To deliver effective diagnostics and therapeutics, understanding the molecular basis of the disease is essential. Contemporary large genome-wide association studies (GWAS) have identified over seventy novel genetic susceptibility loci for LOAD. Most are implicated in microglial or inflammatory pathways, bringing inflammation to the fore as a candidate pathological pathway. Among the most significant GWAS hits are three complement genes: *CLU*, encoding the fluid-phase complement inhibitor clusterin; *CR1* encoding complement receptor 1 (CR1); and recently, *C1S* encoding the complement enzyme C1s. Complement activation is a critical driver of inflammation; changes in complement genes may impact risk by altering the inflammatory status in the brain. To assess complement gene association with LOAD risk, we manually created a comprehensive complement gene list and tested these in gene-set analysis with LOAD summary statistics. We confirmed associations of *CLU* and *CR1* genes with LOAD but showed no significant associations for the complement gene-set when excluding *CLU* and *CR1*. No significant association with other complement genes, including *C1S*, was seen in the IGAP dataset; however, these may emerge from larger datasets.

Keywords: complement; complement receptor 1; clusterin; late-onset Alzheimer's disease; genetics; neuroinflammation

1. Introduction

Alzheimer's disease (AD) is the most common cause of dementia in the elderly. Pathologically, AD is a chronic neurodegenerative disease underpinned by neuronal and synaptic loss, the accumulation of amyloid-β plaques, and neurofibrillary tangles composed of hyperphosphorylated tau. An important role for neuroinflammation has emerged in recent years. Evidence includes the presence of activated microglia in the brain innate immune cells, the presence of inflammatory markers, including complement proteins, in the brain, cerebrospinal fluid (CSF) and plasma, and the demonstration that chronic use of anti-inflammatory drugs may reduce disease incidence [1–3]. Perhaps the best evidence that inflammation may be involved in AD aetiology comes from genome-wide association studies (GWAS); many of the genes most strongly associated with AD risk are involved in inflammation and immunity.

The first causative mutations for AD, identified over 25 years ago in the rare early-onset familial forms of AD, were in Amyloid precursor protein (*APP*), Presenilin 1 (*PSEN1*) and Presenilin 2 (*PSEN2*) genes [4–6]. APP, encoded by the *APP* gene, a broadly expressed transmembrane protein abundant in the brain, is sequentially cleaved by secretase enzymes. The precise cleavage patterns determine its propensity to seed Aβ plaques. The presenilin proteins PSEN1 and PSEN2 are both components of the γ-secretase complex and important in the function of this enzyme; mutations in the genes encoding these proteins impact the APP cleavage pathway. The identification of early-onset AD-associated mutations in these three genes underpins the amyloid cascade hypothesis whereby abnormal APP processing leading to Aβ plaque formation is considered the key underlying pathology associated with AD [7]. However, it is important to stress that these mutations are only relevant to early-onset familial AD which accounts for fewer than 1% of all AD cases. In late-onset Alzheimer's disease (LOAD), accounting for the large majority of AD cases, the strongest genetic risk factor is the presence of the ε4 allele of the gene encoding Apolipoprotein E (ApoE); ε4 confers increased risk, while the most common allele, ε3, is considered neutral for AD, and ε2 has a minor protective effect [8–11]. Homozygosity for *APOE* ε4 confers an ~11-fold increased risk of LOAD compared to ε3 homozygotes. Precisely how these variants in *APOE* impact disease risk remains a subject of ongoing research. ApoE is a lipoprotein present in biological fluids; therefore, roles in lipid transport and membrane repair in the brain have been proposed [12].

Over the past decade, large GWAS have identified variants in more than 70 genetic loci that are associated with LOAD, implicating multiple and diverse biological pathways [13–16]. Notably, ~20% of the genes in LOAD risk loci encode proteins with roles in inflammation and immunity [14,17,18]; many of these are predominantly expressed in microglia, notably *TREM2*, *ABI3* and *PLCG2* [15,19]. From GWAS, it has been shown that three complement system genes are significantly associated with LOAD: *CLU*, *CR1*, and recently, *C1S* encoding the classical pathway enzyme C1s was added to this list [13,16,20]. *CLU* encodes clusterin, a multifunctional plasma protein that regulates the complement terminal pathway, and *CR1* encodes complement receptor 1 (CR1), a receptor for complement fragments and regulator of activation. These are both regulators of the complement cascade and provide the impetus for this analysis of complement genetics in LOAD. To test whether complement genes beyond *CLU* and *CR1* (both genome-wide significant (GWS) in the International Genomics of Alzheimer's Project (IGAP) dataset) influence the risk of LOAD, we compiled a comprehensive complement gene-set containing only those genes that encoded proteins directly involved in complement activation, regulation, or recognition. Then, we undertook several methods of pathway analysis to test whether additional genes within the complement gene-set were associated with LOAD risk.

2. Materials and Methods

2.1. Complement Genes and Gene Exclusion Analyses in LOAD

In order to understand the genetics of the complement pathway in AD, we compiled a comprehensive gene-set comprising all complement genes and associated regulators and receptors. Genes were selected for inclusion based upon known biological relevance to the complement system rather than by using often inaccurate annotations in public databases. The resultant complement gene-set contained 56 genes, subdivided into their relevant functional groups (Table 1).

Table 1. Complement gene list including all complement genes and associated regulators and receptors. Genes are subdivided according to pathway; either classical, lectin, amplification loop or terminal and whether they are complement genes or associated regulators/receptors.

Pathway	HGNC Gene Name	Entrez Gene ID	HGNC Full Gene Name
Classical	C1QA	712	complement C1q A chain
Classical	C1QB	713	complement C1q B chain
Classical	C1QC	714	complement C1q C chain
Classical	C1R	715	complement C1r
Classical	C1S	716	complement C1s
Classical/Lectin	C2	717	complement C2
Classical/Lectin	C4A	720	complement C4A (Rodgers blood group)
Classical/Lectin	C4B	721	complement C4B (Chido blood group)
Lectin	FCN1	2219	ficolin 1
Lectin	FCN2	2220	ficolin 2
Lectin	FCN3	8547	ficolin 3
Lectin	MASP1	5648	mannan binding lectin serine peptidase 1
Lectin	MASP2	10747	mannan binding lectin serine peptidase 2
Lectin	MBL2	4153	mannose binding lectin 2
Amplification loop	CFB	629	complement factor B
Amplification loop	CFD	1675	complement factor D
Classical/Lectin/ Amplification loop	C3	718	complement C3
Terminal	C5	727	complement C5
Terminal	C6	729	complement C6
Terminal	C7	730	complement C7
Terminal	C8A	731	complement C8 α chain
Terminal	C8B	732	complement C8 β chain
Terminal	C8G	733	complement C8 γ chain
Terminal	C9	735	complement C9
Regulator/Receptor	C1QBP	708	complement C1q binding protein
Regulator/Receptor	C3AR1	719	complement C3a receptor 1
Regulator/Receptor	C4BPA	722	complement component 4 binding protein α
Regulator/Receptor	C4BPB	725	complement component 4 binding protein β
Regulator/Receptor	C5AR1	728	complement C5a receptor 1
Regulator/Receptor	C5AR2	27202	complement component 5a receptor 2
Regulator/Receptor	CD46	4179	CD46 molecule
Regulator/Receptor	CD55	1604	CD55 molecule (Cromer blood group)
Regulator/Receptor	CD59	966	CD59 molecule
Regulator/Receptor	CFH	3075	complement factor H
Regulator/Receptor	CFHR1	3078	complement factor H related 1
Regulator/Receptor	CFHR2	3080	complement factor H related 2
Regulator/Receptor	CFHR3	10878	complement factor H related 3
Regulator/Receptor	CFHR4	10877	complement factor H related 4
Regulator/Receptor	CFHR5	81494	complement factor H related 5
Regulator/Receptor	CFI	3426	complement factor I
Regulator/Receptor	CFP	5199	complement factor properdin
Regulator/Receptor	CLU	1191	clusterin
Regulator/Receptor	CR1	1378	complement C3b/C4b receptor 1 (Knops blood group)
Regulator/Receptor	CR2	1380	complement C3d receptor 2
Regulator/Receptor	CSMD1	64478	CUB and Sushi multiple domains 1
Regulator/Receptor	ITGAM	3684	integrin subunit α M
Regulator/Receptor	ITGAX	3687	integrin subunit α X
Regulator/Receptor	SERPING1	710	serpin family G member 1
Regulator/Receptor	VTN	7448	Vitronectin
Regulator/Receptor	CD93	22918	C1q receptor phagocytosis
Complement-like	C1QL1	10882	complement C1q-like 1
Complement-like	C1QL2	165257	complement C1q-like 2
Complement-like	C1QL3	389941	complement C1q-like 3
Complement-like	C1QL4	338761	complement C1q-like 4
Complement-like	C1RL	51279	complement C1r subcomponent-like
Complement-like	CR1L	1379	complement C3b/C4b receptor 1-like

2.2. AD Summary Statistics

This study utilised summary statistics from the International Genomics of Alzheimer's Project (IGAP). IGAP is a large three-stage study based upon GWAS on individuals of European ancestry. In stage 1, IGAP used genotyped and imputed data on 11,480,632 single nucleotide polymorphisms (SNPs) to meta-analyse GWAS datasets consisting of 21,982 Alzheimer's disease cases and 41,944 cognitively normal controls from four consortia: the Alzheimer Disease Genetics Consortium (ADGC); the European Alzheimer's disease Initiative (EADI); the Cohorts for Heart and Aging Research in Genomic Epidemiology Consortium (CHARGE); and the Genetic and Environmental Risk in AD Consortium Genetic and Environmental Risk in AD/Defining Genetic, Polygenic and Environmental Risk for Alzheimer's Disease Consortium (GERAD/PERADES). In stage 2, 11,632 SNPs were genotyped and tested for association in an independent set of 8362 Alzheimer's disease cases and 10,483 controls. Meta-analyses of variants selected for analysis in stage 3A (n = 11,666) or stage 3B (n = 30,511) samples brought the final sample to 35,274 clinical and autopsy-documented Alzheimer's disease cases and 59,163 controls.

Gene-set analysis was performed using the complement gene-set and stage 1 summary statistics from the International Genomics of Alzheimer's Project [14]. The individual and combined effects of the genome-wide significant (GWS) genes *CLU* and *CR1* within the complement gene-set were investigated by removing these genes individually and together. We utilised the most up-to-date publicly available GWAS dataset at the time of writing [14], and calculated the complement gene-set *p*-values when including and excluding those loci that reached genome-wide significance in the IGAP dataset. The recently identified LOAD-associated *C1S* variant [13] does not show genome-wide statistical significance in the IGAP dataset; and therefore was not removed in the gene-set analysis. Complement gene-sets were tested for enrichment using the IGAP stage 1 summary statistics [14] in MAGMA version 1.06 [21]. Summary statistics were filtered for common variants (MAF $\geq$ 0.01) and all indels and merged deletions were removed; 8,608,484 SNPs were analysed. Genes were annotated using reference data files from the European population of Phase 3 of 1000 Genomes, human genome Build 37 using a window of 35 kb upstream and 10 kb downstream of each gene [22]. Ten thousand permutations were used to estimate *p*-values, corrected for multiple testing using the family-wise error rate (FWER). Gene-sets with a FWER-corrected *p*-value < 0.05 under the "mean" model for estimating gene-level associations were reported as significant.

2.3. Complement Risk Score Analysis

A complement risk score combining the effects of all SNPs in the complement gene-set was produced. POLARIS [23] was used to compute risk scores in GERAD-genotyped data (3332 cases, 9832 controls) using SNP effect sizes from IGAP stage 1 summary statistics [14,16,20] (excluding GERAD subjects). Linkage disequilibrium (LD) was estimated from the GERAD data, and POLARIS was used to adjust the scores for LD between SNPs. The overall association of the complement gene-set with LOAD was determined using a logistic regression model, adjusting for population covariates, age, and sex. The logistic regression model included the baseline polygenic risk scores for all SNPs in the model, thereby testing for any association beyond the baseline polygenic effect.

Data used in the preparation of this article were obtained from the Genetic and Environmental Risk for Alzheimer's disease (GERAD) Consortium. The imputed GERAD sample comprised 3177 AD cases and 7277 controls with available age and gender data. Cases and elderly screened controls were recruited by the Medical Research Council (MRC) Genetic Resource for AD (Cardiff University; Institute of Psychiatry, London; Cambridge University; Trinity College Dublin), the Alzheimer's Research UK (ARUK) Collaboration (University of Nottingham; University of Manchester; University of Southampton; University of Bristol; Queen's University Belfast; the Oxford Project to Investigate Memory and Ageing (OPTIMA), Oxford University); Washington University, St Louis, United States; MRC PRION Unit, University College London; London and the South East Region

AD project (LASER-AD), University College London; Competence Network of Dementia (CND) and Department of Psychiatry, University of Bonn, Germany; the National Institute of Mental Health (NIMH) AD Genetics Initiative. A total of 6129 population controls were drawn from large existing cohorts with available GWAS data, including the 1958 British Birth Cohort (1958BC) (http://www.b58cgene.sgul.ac.uk, accessed on 15 March 2021), the KORA F4 Study, and the Heinz Nixdorf Recall Study. All AD cases met criteria for either probable (NINCDS-ADRDA, DSM-IV) or definite (CERAD) AD. All elderly controls were screened for dementia using the MMSE or ADAS-cog and were determined to be free from dementia at neuropathological examination or had a Braak score of 2.5 or lower. Genotypes from all cases and 4617 controls were previously included in the AD GWAS by Harold and colleagues (2009) [20]. Genotypes for the remaining population controls were obtained from WTCCC2. Imputation of the dataset was performed using IMPUTE2 and the 1000 genomes (http://www.1000genomes.org/, accessed on 15 March 2021) Dec2010 reference panel (NCBI build 37.1).

2.4. Likelihood Ratio Analysis

A likelihood ratio test was used to estimate how much of the complement gene-set effect on LOAD risk was contributed by *CLU* and *CR1*, and to test whether there were residual polygenic effects of the remaining genes from the complement gene-set. The effects of *CLU* and *CR1* were estimated using a risk score combining all SNPs in the gene, produced using POLARIS in order to correct for LD. Likelihood ratio tests were used to compare individual models containing SNPs in *CLU* and *CR1* and models containing the combined risk conferred by SNPs in the rest of the complement gene-set.

3. Results

3.1. MAGMA Analysis Reveals the Impact of Individual Complement Genes

From the MAGMA gene-set analysis, the complement gene-set comprising all 56 genes was significantly associated with LOAD ($p = 0.011$) (Table 2). When the GWAS-significant genes *CLU* and *CR1* were excluded individually from the gene-set, the complement-minus-*CLU* gene-set was not significant ($p = 0.057$), while the complement-minus-*CR1* gene-set was significant ($p = 0.048$). As *CR1* and *CR1L* are located next to each other on chromosome 1, and linkage disequilibrium extends between the two genes, we excluded the *CR1/CR1L* locus from the gene-set. This gene-set was not significant ($p = 0.082$). The gene set in which both *CLU* and *CR1* were excluded from the complement gene-set was not significantly associated with LOAD ($p = 0.170$). The signal in the gene-set where *CR1L*, *CLU* and *CR1* were excluded was reduced compared with the signal derived from the gene-sets in which *CLU* and *CR1* were removed (Table 2). Taken together, these results suggest that the LOAD association signal in the complement gene-set is predominantly driven by *CLU* and *CR1*. Given the physical distance between *CR1* and *CR1L*, the use of extended gene boundaries and that linkage disequilibrium extends across both genes, we cannot resolve the signal between these two genes in the gene set analysis. Hence, we cannot confirm any independent contribution from *CR1L*.

Table 2. Complement gene-set analysis.

Gene-Set	Ngenes	OR	95% CI	p	p FWER
Complement Genes	56	1.402	[1.068, 1.841]	0.008	0.011
Complement Genes Minus *CLU*	55	1.278	[0.969, 1.684]	0.041	0.057
Complement Genes Minus *CR1*	55	1.288	[0.981, 1.691]	0.034	0.048
Complement Genes Minus *CLU*, *CR1*	54	1.172	[0.891, 1.542]	0.129	0.170
Complement Genes Minus *CR1*, *CR1L*	54	1.244	[0.943, 1.639]	0.061	0.082
Complement Genes Minus *CLU*, *CR1*, *CR1L*	53	1.127	[0.854, 1.489]	0.199	0.246

Table 2 displays the results from the MAGMA analysis. Gene-sets were corrected for multiple testing using the family-wise error rate (FWER). The complement gene-set is significant ($p = 0.011$), but this effect is lost when CLU and CR1 are excluded from the gene-set ($p = 0.170$). CLU has the largest impact in the complement set, and the association with AD is predominantly driven by CLU and CR1.

3.2. Risk Score Analysis Supports the Impact of Complement Genes

To further explore the impact of complement genes on LOAD risk, we adopted a polygenic approach. We first applied risk score analysis to the dataset, then used logistic regression to explore the association between LOAD and complement gene-set risk scores in GERAD individuals (Table 3). The complement gene-set as a whole was strongly associated with AD in this analysis ($p = 0.003$). Removal of CLU from the gene-set caused the largest reduction in significance ($p = 0.003$ vs. $p = 0.053$). Removal of CR1, or the CR1/CR1L locus had minimal impact on the significance of association in the gene-set, although when CLU, CR1 and CR1L were eliminated, the significance was further reduced compared to the elimination of CLU alone ($p = 0.148$ vs. $p = 0.053$) (Table 3). These gene elimination analyses demonstrated that CLU and CR1 were the major contributors to the risk of LOAD in the complement gene-set; however, the polygenic approach revealed that CLU was by far the more significant of these. In these data, the CLU gene shows a stronger association compared to CR1 ($p = 1.03 \times 10^{-5}$ and $p = 1.5 \times 10^{-3}$, respectively). The joint association of CLU and CR1 is stronger still ($p = 3.88 \times 10^{-7}$), showing that CLU and CR1 are both independently associated with AD.

Table 3. Association between Alzheimer's disease (AD) and complement gene-set risk score.

Gene-Set	Ngenes	OR	95% CI	p
Complement Genes	56	1.090	[1.028, 1.156]	0.003
Complement Genes Minus CLU	55	1.059	[0.998, 1.123]	0.053
Complement Genes Minus CR1	55	1.089	[1.027, 1.155]	0.004
Complement Genes Minus CLU, CR1	54	1.058	[0.997, 1.122]	0.059
Complement Genes Minus CR1, CR1L	54	1.077	[1.015, 1.142]	0.013
Complement Genes Minus CLU, CR1, CR1L	53	1.044	[0.984, 1.107]	0.148

Table 3 displays the results from the risk score analysis; the overall complement risk score shows an association with AD ($p = 0.003$). CLU explains the majority of this signal.

3.3. Likelihood Ratio Analysis Confirms No Significant Impact of Other Complement Genes

We next tested complement gene-set effects using likelihood ratio analyses. Models in which CLU, CR1 and CR1/CR1L were removed individually, showed significant residual impact in the gene-set ($p = 0.0136$; $p = 0.0091$; $p = 0.0063$ respectively); after removal of CR1 and CLU or CR1, CLU and CR1L, there was no significant residual impact in the gene-set, demonstrating that there was no significant polygenic effect of the remaining complement genes in the datasets used (Table 4). These results further support the conclusion that the complement gene-set association with LOAD is driven predominantly by CLU and CR1, but with no significant contribution from other complement gene-set members ($p = 0.1457$; Table 4).

Table 4 shows the results from these likelihood ratio tests comparing models containing SNPs in CLU, CR1 and CR1/CR1L only and models containing the combined risk in SNPs in the remaining complement genes. The p-values demonstrate whether the remaining genes in the complement explain any additional variation. These results further support the conclusion that the complement gene-set impact on LOAD risk is predominantly driven by CLU and CR1.

Table 4. Likelihood ratio test (LRT) comparing gene-set risk scores.

Models Compared	LRT p-Value
(1) *CLU* (2) *CLU* + Complement_minus_*CLU*	0.0136
(1) *CR1* (2) *CR1* + Complement_minus_*CR1*	0.0091
(1) *CLU* + *CR1* (2) *CLU* + *CR1* + Complement_minus_*CLU_CR1*	0.1457
(1) *CR1* + *CR1L* (2) *CR1* + *CR1L* + Complement_minus_*CR1_CR1L*	0.0063
(1) *CLU* + *CR1* + *CR1L* (2) *CLU* + *CR1* + *CR1L* + Complement_minus_*CLU_CR1_CR1L*	0.1145

4. Discussion

The first evidence implicating the complement system in LOAD came from immunostaining of post-mortem brain tissue. Complement components and activation products, notably C1q, C4b, C3b/iC3b and the membrane attack complex, were present and co-localised with amyloid plaques and neurofibrillary tangles in the AD brain [24–27]. C3 fragments were shown to opsonise amyloid for phagocytosis by microglia in the brain and facilitate transport on erythrocytes to the liver [28]. Complement activation is critically involved in synaptic pruning both in development and in diseases such as AD [29–32]. In AD mouse models, back-crossing to complement deficiencies has supported the critical role of complements in neuroinflammation and synapse loss [30,33]. The presence of complement activation biomarkers in CSF and/or plasma in LOAD suggested that complement dysregulation occurs early in the disease [2]. The demonstration that complement genes associated with LOAD provided compelling evidence that the complement was a driver of disease rather than a secondary event [13,14,16].

To further investigate the roles of complement genes in the risk of LOAD, we compiled a comprehensive complement gene-set and used a polygenic approach to identify genes contributing to AD risk. We have demonstrated that the signal for the association of the complement gene-set with LOAD is explained by the GWS genes *CLU* and *CR1*, and not by other complement genes tested here. This finding was unexpected. Based on knowledge from other chronic inflammatory diseases, we had hypothesised that many complement genes might influence LOAD risk. For example, in age-related macular degeneration (AMD), a retinal disease clinically and pathologically linked to LOAD, genes encoding complement components C2, C3, FB and C9, and regulators FH and FHR4, all contribute to risk [34]. Indeed, the demonstration that multiple complement genes can collaborate to cause dysregulation and disease informed the concept of the "complotype", the set of complement gene variants inherited by an individual that dictates complement activity and disease risk [35]. The genetic associations in these other chronic inflammatory diseases influence systemic or local complement regulation and/or amplification of activation; these in turn cause complement dysregulation that drives inflammation. Our demonstration that the complement genetic signature in LOAD is restricted to the genes encoding clusterin and CR1 suggests that complement dysregulation is not critical in the disease process. However, it should be noted that this finding is dependent on the dataset being investigated. At the time of writing, we utilised the largest publicly available AD GWAS dataset [14]. A recent study by the European AD Biobank, currently a preprint, reported an LOAD GWAS-significant association with the complement gene *C1S* [13]; this suggests that larger datasets and different analytical methods may implicate other complement genes and further elucidate roles of the complement system in LOAD. Additionally, because of the highly repetitive nature of a number of the complement loci, for example, the regulators of complement activation (RCA) clusters on chromosome 1 [36], many complement genes may be hidden from standard sequencing technologies; the application of emerging long-range sequencing methods may reveal additional genetic variation in complement genes linked to LOAD missed in current GWAS and whole exome/genome sequencing studies using short read sequencing technologies [37].

Of the complement genes tested here, *CLU* and *CR1* were significantly associated with LOAD through multiple analytical approaches. Clusterin is a multi-functional plasma protein; its role in the complement system is to restrict fluid-phase membrane attack pathway activation [38]; however, beyond the complement system, clusterin functions as an extracellular chaperone protein, is involved in oxidative stress and cell survival/cell death pathways, and functions as an apolipoprotein in lipid transport [38–41]. Any one or several of these functions might underpin the association with LOAD. Four SNPs in *CLU*, all intronic and in LD, have been associated with increased LOAD risk (rs11136000, rs2279590, rs9331888 and rs9331896) [16,20]; evidence to date suggests that these SNPs impact clusterin synthesis, and hence, plasma clusterin levels. CR1 is a membrane-bound receptor for complement components (C1q, MBL) and fragments (C3b, C4b). The primary function of CR1 is as a receptor for C3b/C4b-opsonised immune complexes. CR1 on erythrocytes sequesters immune complexes and transports them to disposal sites, while CR1 on phagocytic cells binds opsonised immune complexes and processes them for elimination via phagocytosis. This latter activity requires a second function of CR1, its cofactor activity for factor I cleavage of C3b to iC3b the ligand for the phagocytic receptor CR3. The biological relevance of the C1q/MBL binding functions of CR1 are unclear. The human *CR1* gene is located in the RCA gene cluster on chromosome 1 (1q32); duplications and deletions in this highly repetitive gene generate multiple isoforms via copy number variation (CNV). The most common variant, CR1*1 (allele frequency 0.87) comprises 30 tandem repeats of 60–70 amino acid units called short consensus repeats (SCRs), which are in turn grouped in four homologous sets of seven termed long homologous repeats (LHRs), each a separate C3b/C4b binding unit. The second most common variant CR1*2 (allele frequency 0.11) is identical to CR1*1 except for the acquisition of an additional LHR, a "gain-of-function"; this variant increases risk for LOAD by up to 30%, although precisely how is unclear [14,16,42–44]. It has been suggested that the CR1*2 variant is associated with lower CR1 expression on erythrocytes, reducing the efficiency of peripheral immune complex handling and impacting amyloid clearance from the brain [45,46].

Our original analysis suggested that some of the signal from the complement gene-set might be attributable to the *CR1L* gene. However, *CR1L* is immediately adjacent to *CR1* and the SNP signals cannot be resolved, so it is not possible to ascribe an independent signal to *CR1L* in this analysis. *CR1L* encodes a C4b-binding protein comprising 13 SCRs, expressed predominantly in haematopoietic tissues [47,48]. Its physiological role is unknown, and evidence mechanistically linking it to LOAD is absent.

5. Conclusions

Taken together, our findings confirm the strong genetic association of the complement genes *CR1* and *CLU* with LOAD and that there is no statistically significant association signal for other complement genes apparent in the dataset used for the analysis. CR1 and clusterin are important regulators of the complement pathway, suggesting that its dysregulation is important in LOAD. The recent GWAS association of *C1S* with LOAD demonstrates the potential for missing associations in this complex gene-set and raises the possibility that other loci may be missed by current large-scale genotyping and short-read sequencing technologies. Application of long read sequencing technologies could significantly alter the current landscape of complement system genetics in relation to LOAD risk.

Supplementary Materials: The following are available online at https://www.mdpi.com/2073-442 5/12/3/443/s1, Supplementary File 1: List of GERAD authors and affiliations.

Author Contributions: Conceptualization, B.P.M.; data curation, S.M.C., J.C.H., E.B., M.T. and The GERAD1 Consortium; formal analysis, J.C.H. and E.B.; funding acquisition, J.W. and B.P.M.; investigation, J.C.H. and E.B.; methodology, J.C.H. and E.B.; software, J.C.H. and E.B.; supervision, B.P.M.; writing—original draft preparation, S.M.C., J.C.H., E.B., M.T. and B.P.M.; writing—review and editing, S.M.C., J.C.H., E.B., M.T., R.S., J.W. and B.P.M. All authors have read and agreed to the published version of the manuscript.

Funding: This work is supported by the U.K. Dementia Research Institute which receives its funding from U.K. DRI Ltd., funded by the U.K. Medical Research Council, Alzheimer's Society, and Alzheimer's Research U.K.

Data Availability Statement: The GERAD data is available by request (GERADConsortium@cf.ac.uk) and Kunkle data is available to download (https://www.niagads.org/datasets/ng00075 accessed on 15 March 2021).

Acknowledgments: Data used in the preparation of this article were obtained from the Genetic and Environmental Risk for Alzheimer's disease (GERAD1) Consortium. Cardiff University was supported by the Wellcome Trust, Medical Research Council (MRC), Alzheimer's Research U.K. (ARUK) and the Welsh Assembly Government. Cambridge University and Kings College London acknowledge support from the MRC. ARUK supported sample collections at the South West Dementia Bank and the Universities of Nottingham, Manchester, and Belfast. The Belfast group acknowledges support from the Alzheimer's Society, Ulster Garden Villages, Northern Ireland R&D Office, and the Royal College of Physicians/Dunhill Medical Trust. The MRC and Mercer's Institute for Research on Ageing supported the Trinity College group. The South West Dementia Brain Bank acknowledges support from Bristol Research into Alzheimer's and Care of the Elderly. The Charles Wolfson Charitable Trust supported the OPTIMA group. Washington University was funded by NIH grants, Barnes Jewish Foundation and the Charles and Joanne Knight Alzheimer's Research Initiative. Patient recruitment for the MRC Prion Unit/UCL Department of Neurodegenerative Disease collection was supported by the UCLH/UCL Biomedical Centre and NIHR Queen Square Dementia Biomedical Research Unit. LASER-AD was funded by Lundbeck SA. The Bonn group was supported by the German Federal Ministry of Education and Research (BMBF), Competence Network Dementia and Competence Network Degenerative Dementia, and by the Alfried Krupp von Bohlen und Halbach-Stiftung. The GERAD1 Consortium also used samples ascertained by the NIMH AD Genetics Initiative. The GERAD data used in this paper included 5770 additional population controls; including the 1958 British Birth Cohort (1958BC), the KORA F4 Study, Heinz Nixdorf Recall Study and controls from the National Blood Service genotyped as part of the Wellcome Trust Case Control Consortium. The KORA F4 studies were financed by Helmholtz Zentrum München; German Research Center for Environmental Health; BMBF; German National Genome Research Network and the Munich Center of Health Sciences. The Heinz Nixdorf Recall cohort was funded by the Heinz Nixdorf Foundation (jur. G. Schmidt, Chairman) and BMBF. Coriell Cell Repositories is supported by NINDS and the Intramural Research Program of the National Institute on Aging. We acknowledge the use of genotype data from the 1958 Birth Cohort collection, funded by the MRC and the Wellcome Trust, which was genotyped by the Wellcome Trust Case Control Consortium and the Type-1 Diabetes Genetics Consortium, sponsored by the National Institute of Diabetes and Digestive and Kidney Diseases, National Institute of Allergy and Infectious Diseases, National Human Genome Research Institute, National Institute of Child Health and Human Development and Juvenile Diabetes Research Foundation International. We thank the International Genomics of Alzheimer's Project for providing summary data results for these analyses. The investigators within IGAP contributed to the design and implementation of IGAP and/or provided data but did not participate in the analysis or writing of this report. IGAP was made possible by the generous participation of the control subjects, the patients, and their families. The i-Select chips were funded by the French National Foundation on Alzheimer's disease and related disorders. EADI was supported by the LABEX (laboratory of excellence program investment for the future) DISTALZ grant, Inserm, Institut Pasteur de Lille, Université de Lille 2, and the Lille University Hospital. GERAD/PERADES was supported by the Medical Research Council (grant no. 503480), Alzheimer's Research U.K. (grant no. 503176), the Wellcome Trust (grant no. 082604/2/07/Z) and the German Federal Ministry of Education and Research (BMBF): Competence Network Dementia (CND) grant no. 01GI0102, 01GI0711, 01GI0420.

CHARGE was partly supported by the NIH/NIA grant R01 AG033193 and the NIA AG081220 and AGES contract N01–AG–12100, the NHLBI grant R01 HL105756, the Icelandic Heart Association, and the Erasmus Medical Center and Erasmus University. ADGC was supported by the NIH/NIA grants: U01 AG032984, U24 AG021886, U01 AG016976, and the Alzheimer's Association grant ADGC–10–196728.

Conflicts of Interest: The authors declare no conflict of interest. The funders had no role in the design of the study; in the collection, analyses, or interpretation of data; in the writing of the manuscript, or in the decision to publish the results.

References

1. Etminan, M.; Gill, S.; Samii, A. Effect of Non-Steroidal Anti-Inflammatory Drugs on Risk of Alzheimer's Disease: Systematic Review and Meta-Analysis of Observational Studies. *BMJ* **2003**, *327*, 128. [CrossRef] [PubMed]
2. Hakobyan, S.; Harding, K.; Aiyaz, M.; Hye, A.; Dobson, R.; Baird, A.; Liu, B.; Harris, C.L.; Lovestone, S.; Morgan, B.P. Complement Biomarkers as Predictors of Disease Progression in Alzheimer's Disease. *J. Alzheimers Dis.* **2016**, *54*, 707–716. [CrossRef]
3. Morgan, A.R.; Touchard, S.; O'Hagan, C.; Sims, R.; Majounie, E.; Escott-Price, V.; Jones, L.; Williams, J.; Morgan, B.P. The Correlation between Inflammatory Biomarkers and Polygenic Risk Score in Alzheimer's Disease. *J. Alzheimers Dis.* **2017**, *56*, 25–36. [CrossRef]
4. Goate, A.; Chartier-Harlin, M.C.; Mullan, M.; Brown, J.; Crawford, F.; Fidani, L.; Giuffra, L.; Haynes, A.; Irving, N.; James, L.; et al. Segregation of a missense mutation in the amyloid precursor protein gene with familial Alzheimer's disease. *Nature* **1991**, *349*, 704–706. [CrossRef] [PubMed]
5. Sherrington, R.; Rogaev, E.I.; Liang, Y.; Rogaeva, E.A.; Levesque, G.; Ikeda, M.; Chi, H.; Lin, C.; Li, G.; Holman, K.; et al. Cloning of a gene bearing missense mutations in early-onset familial Alzheimer's disease. *Nature* **1995**, *375*, 754–760. [CrossRef] [PubMed]
6. Levy-Lahad, E.; Wasco, W.; Poorkaj, P.; Romano, D.M.; Oshima, J.; Pettingell, W.H.; Yu, C.E.; Jondro, P.D.; Schmidt, S.D.; Wang, K.; et al. Candidate gene for the chromosome 1 familial Alzheimer's disease locus. *Science* **1995**, *269*, 973–977. [CrossRef]
7. Hardy, J. The discovery of Alzheimer-causing mutations in the APP gene and the formulation of the "amyloid cascade hypothesis". *FEBS J.* **2017**, *284*, 1040–1044. [CrossRef]
8. Corder, E.H.; Saunders, A.M.; Risch, N.J.; Strittmatter, W.J.; Schmechel, D.E.; Gaskell, P.C., Jr.; Rimmler, J.B.; Locke, P.A.; Connealy, P.M.; Schmader, K.E.; et al. Protective effect of apolipoprotein E type 2 allele for late onset Alzheimer disease. *Nat. Genet.* **1994**, *7*, 180–184. [CrossRef]
9. Saunders, A.M.; Schmader, K.; Breitner, J.C.; Benson, M.D.; Brown, W.T.; Goldfarb, L.; Goldgaber, D.; Manwaring, M.G.; Szymanski, M.H.; McCown, N.; et al. Apolipoprotein E epsilon 4 allele distributions in late-onset Alzheimer's disease and in other amyloid-forming diseases. *Lancet* **1993**, *342*, 710–711. [CrossRef]
10. Saunders, A.M.; Strittmatter, W.J.; Schmechel, D.; George-Hyslop, P.H.; Pericak-Vance, M.A.; Joo, S.H.; Rosi, B.L.; Gusella, J.F.; Crapper-MacLachlan, D.R.; Alberts, M.J.; et al. Association of apolipoprotein E allele epsilon 4 with late-onset familial and sporadic Alzheimer's disease. *Neurology* **1993**, *43*, 1467–1472. [CrossRef]
11. Strittmatter, W.J.; Weisgraber, K.H.; Huang, D.Y.; Dong, L.M.; Salvesen, G.S.; Pericak-Vance, M.; Schmechel, D.; Saunders, A.M.; Goldgaber, D.; Roses, A.D. Binding of human apolipoprotein E to synthetic amyloid beta peptide: Isoform-specific effects and implications for late-onset Alzheimer disease. *Proc. Natl. Acad. Sci. USA* **1993**, *90*, 8098–8102. [CrossRef]
12. Lanfranco, M.F.; Ng, C.A.; Rebeck, G.W. ApoE Lipidation as a Therapeutic Target in Alzheimer's Disease. *Int. J. Mol. Sci.* **2020**, *21*, 6336. [CrossRef]
13. Bellenguez, C.; Küçükali, F.; Jansen, I.; Andrade, V.; Moreno-Grau, S.; Amin, N.; Naj, A.C.; Grenier-Boley, B.; Campos-Martin, R.; Holmans, P.A.; et al. New insights on the genetic etiology of Alzheimer's and related dementia. *medRxiv* **2020**. [CrossRef]
14. Kunkle, B.W.; Grenier-Boley, B.; Sims, R.; Bis, J.C.; Damotte, V.; Naj, A.C.; Boland, A.; Vronskaya, M.; van der Lee, S.J.; Amlie-Wolf, A.; et al. Genetic meta-analysis of diagnosed Alzheimer's disease identifies new risk loci and implicates Abeta, tau, immunity and lipid processing. *Nat. Genet.* **2019**, *51*, 414–430. [CrossRef]
15. Sims, R.; van der Lee, S.J.; Naj, A.C.; Bellenguez, C.; Badarinarayan, N.; Jakobsdottir, J.; Kunkle, B.W.; Boland, A.; Raybould, R.; Bis, J.C.; et al. Rare coding variants in PLCG2, ABI3, and TREM2 implicate microglial-mediated innate immunity in Alzheimer's disease. *Nat. Genet.* **2017**, *49*, 1373–1384. [CrossRef] [PubMed]
16. Lambert, J.C.; Heath, S.; Even, G.; Campion, D.; Sleegers, K.; Hiltunen, M.; Combarros, O.; Zelenika, D.; Bullido, M.J.; Tavernier, B.; et al. Genome-wide association study identifies variants at CLU and CR1 associated with Alzheimer's disease. *Nat. Genet.* **2009**, *41*, 1094–1099. [CrossRef] [PubMed]
17. Jones, L.; Holmans, P.A.; Hamshere, M.L.; Harold, D.; Moskvina, V.; Ivanov, D.; Pocklington, A.; Abraham, R.; Hollingworth, P.; Sims, R.; et al. Genetic evidence implicates the immune system and cholesterol metabolism in the aetiology of Alzheimer's disease. *PLoS ONE* **2010**, *5*, e13950. [CrossRef]
18. Ahmad, S.; Bannister, C.; van der Lee, S.J.; Vojinovic, D.; Adams, H.H.H.; Ramirez, A.; Escott-Price, V.; Sims, R.; Baker, E.; Williams, J.; et al. Disentangling the biological pathways involved in early features of Alzheimer's disease in the Rotterdam Study. *Alzheimers Dement.* **2018**, *14*, 848–857. [CrossRef]

19. Guerreiro, R.; Wojtas, A.; Bras, J.; Carrasquillo, M.; Rogaeva, E.; Majounie, E.; Cruchaga, C.; Sassi, C.; Kauwe, J.S.; Younkin, S.; et al. TREM2 variants in Alzheimer's disease. *N. Engl. J. Med.* **2013**, *368*, 117–127. [CrossRef]
20. Harold, D.; Abraham, R.; Hollingworth, P.; Sims, R.; Gerrish, A.; Hamshere, M.L.; Pahwa, J.S.; Moskvina, V.; Dowzell, K.; Williams, A.; et al. Genome-wide association study identifies variants at CLU and PICALM associated with Alzheimer's disease. *Nat. Genet.* **2009**, *41*, 1088–1093. [CrossRef] [PubMed]
21. De Leeuw, C.A.; Mooij, J.M.; Heskes, T.; Posthuma, D. MAGMA: Generalized gene-set analysis of GWAS data. *PLoS Comput. Biol.* **2015**, *11*, e1004219. [CrossRef]
22. Genomes Project, C.; Auton, A.; Brooks, L.D.; Durbin, R.M.; Garrison, E.P.; Kang, H.M.; Korbel, J.O.; Marchini, J.L.; McCarthy, S.; McVean, G.A.; et al. A global reference for human genetic variation. *Nature* **2015**, *526*, 68–74. [CrossRef]
23. Baker, E.; Schmidt, K.M.; Sims, R.; O'Donovan, M.C.; Williams, J.; Holmans, P.; Escott-Price, V.; Consortium, W.T.G. POLARIS: Polygenic LD-adjusted risk score approach for set-based analysis of GWAS data. *Genet. Epidemiol.* **2018**, *42*, 366–377. [CrossRef]
24. Veerhuis, R.; van der Valk, P.; Janssen, I.; Zhan, S.S.; Van Nostrand, W.E.; Eikelenboom, P. Complement activation in amyloid plaques in Alzheimer's disease brains does not proceed further than C3. *Virchows Arch.* **1995**, *426*, 603–610. [CrossRef]
25. Rogers, J.; Cooper, N.R.; Webster, S.; Schultz, J.; McGeer, P.L.; Styren, S.D.; Civin, W.H.; Brachova, L.; Bradt, B.; Ward, P.; et al. Complement activation by beta-amyloid in Alzheimer disease. *Proc. Natl. Acad. Sci. USA* **1992**, *89*, 10016–10020. [CrossRef] [PubMed]
26. Webster, S.; Lue, L.F.; Brachova, L.; Tenner, A.J.; McGeer, P.L.; Terai, K.; Walker, D.G.; Bradt, B.; Cooper, N.R.; Rogers, J. Molecular and cellular characterization of the membrane attack complex, C5b-9, in Alzheimer's disease. *Neurobiol. Aging* **1997**, *18*, 415–421. [CrossRef]
27. Ishii, T.; Haga, S. Immuno-electron-microscopic localization of complements in amyloid fibrils of senile plaques. *Acta Neuropathol.* **1984**, *63*, 296–300. [CrossRef] [PubMed]
28. Bradt, B.M.; Kolb, W.P.; Cooper, N.R. Complement-dependent proinflammatory properties of the Alzheimer's disease beta-peptide. *J. Exp. Med.* **1998**, *188*, 431–438. [CrossRef]
29. Chu, Y.; Jin, X.; Parada, I.; Pesic, A.; Stevens, B.; Barres, B.; Prince, D.A. Enhanced synaptic connectivity and epilepsy in C1q knockout mice. *Proc. Natl. Acad. Sci. USA* **2010**, *107*, 7975–7980. [CrossRef]
30. Hong, S.; Beja-Glasser, V.F.; Nfonoyim, B.M.; Frouin, A.; Li, S.; Ramakrishnan, S.; Merry, K.M.; Shi, Q.; Rosenthal, A.; Barres, B.A.; et al. Complement and microglia mediate early synapse loss in Alzheimer mouse models. *Science* **2016**, *352*, 712–716. [CrossRef] [PubMed]
31. Sekar, A.; Bialas, A.R.; de Rivera, H.; Davis, A.; Hammond, T.R.; Kamitaki, N.; Tooley, K.; Presumey, J.; Baum, M.; Van Doren, V.; et al. Schizophrenia risk from complex variation of complement component 4. *Nature* **2016**, *530*, 177–183. [CrossRef]
32. Stevens, B.; Allen, N.J.; Vazquez, L.E.; Howell, G.R.; Christopherson, K.S.; Nouri, N.; Micheva, K.D.; Mehalow, A.K.; Huberman, A.D.; Stafford, B.; et al. The classical complement cascade mediates CNS synapse elimination. *Cell* **2007**, *131*, 1164–1178. [CrossRef]
33. Shi, Q.; Colodner, K.J.; Matousek, S.B.; Merry, K.; Hong, S.; Kenison, J.E.; Frost, J.L.; Le, K.X.; Li, S.; Dodart, J.C.; et al. Complement C3-Deficient Mice Fail to Display Age-Related Hippocampal Decline. *J. Neurosci.* **2015**, *35*, 13029–13042. [CrossRef]
34. Geerlings, M.J.; de Jong, E.K.; den Hollander, A.I. The complement system in age-related macular degeneration: A review of rare genetic variants and implications for personalized treatment. *Mol. Immunol.* **2017**, *84*, 65–76. [CrossRef] [PubMed]
35. Harris, C.L.; Heurich, M.; Rodriguez de Cordoba, S.; Morgan, B.P. The complotype: Dictating risk for inflammation and infection. *Trends Immunol.* **2012**, *33*, 513–521. [CrossRef]
36. Hourcade, D.; Holers, V.M.; Atkinson, J.P. The regulators of complement activation (RCA) gene cluster. *Adv. Immunol.* **1989**, *45*, 381–416. [CrossRef]
37. Ebbert, M.T.W.; Jensen, T.D.; Jansen-West, K.; Sens, J.P.; Reddy, J.S.; Ridge, P.G.; Kauwe, J.S.K.; Belzil, V.; Pregent, L.; Carrasquillo, M.M.; et al. Systematic analysis of dark and camouflaged genes reveals disease-relevant genes hiding in plain sight. *Genome Biol.* **2019**, *20*, 97. [CrossRef]
38. Tschopp, J.; Chonn, A.; Hertig, S.; French, L.E. Clusterin, the human apolipoprotein and complement inhibitor, binds to complement C7, C8 beta, and the b domain of C9. *J. Immunol.* **1993**, *151*, 2159–2165. [PubMed]
39. Burkey, B.F.; Stuart, W.D.; Harmony, J.A. Hepatic apolipoprotein J is secreted as a lipoprotein. *J. Lipid Res.* **1992**, *33*, 1517–1526. [CrossRef]
40. Humphreys, D.T.; Carver, J.A.; Easterbrook-Smith, S.B.; Wilson, M.R. Clusterin has chaperone-like activity similar to that of small heat shock proteins. *J. Biol. Chem.* **1999**, *274*, 6875–6881. [CrossRef]
41. Jenne, D.E.; Tschopp, J. Molecular structure and functional characterization of a human complement cytolysis inhibitor found in blood and seminal plasma: Identity to sulfated glycoprotein 2, a constituent of rat testis fluid. *Proc. Natl. Acad. Sci. USA* **1989**, *86*, 7123–7127. [CrossRef] [PubMed]
42. Brouwers, N.; Van Cauwenberghe, C.; Engelborghs, S.; Lambert, J.C.; Bettens, K.; Le Bastard, N.; Pasquier, F.; Montoya, A.G.; Peeters, K.; Mattheijssens, M.; et al. Alzheimer risk associated with a copy number variation in the complement receptor 1 increasing C3b/C4b binding sites. *Mol. Psychiatry* **2012**, *17*, 223–233. [CrossRef]

43. Hazrati, L.N.; Van Cauwenberghe, C.; Brooks, P.L.; Brouwers, N.; Ghani, M.; Sato, C.; Cruts, M.; Sleegers, K.; St George-Hyslop, P.; Van Broeckhoven, C.; et al. Genetic association of CR1 with Alzheimer's disease: A tentative disease mechanism. *Neurobiol. Aging* **2012**, *33*, e2945–e2949. [CrossRef] [PubMed]
44. Kucukkilic, E.; Brookes, K.; Barber, I.; Guetta-Baranes, T.; Consortium, A.; Morgan, K.; Hollox, E.J. Complement receptor 1 gene (CR1) intragenic duplication and risk of Alzheimer's disease. *Hum. Genet.* **2018**, *137*, 305–314. [CrossRef] [PubMed]
45. Mahmoudi, R.; Feldman, S.; Kisserli, A.; Duret, V.; Tabary, T.; Bertholon, L.A.; Badr, S.; Nonnonhou, V.; Cesar, A.; Neuraz, A.; et al. Inherited and Acquired Decrease in Complement Receptor 1 (CR1) Density on Red Blood Cells Associated with High Levels of Soluble CR1 in Alzheimer's Disease. *Int. J. Mol. Sci.* **2018**, *19*, 2175. [CrossRef]
46. Mahmoudi, R.; Kisserli, A.; Novella, J.L.; Donvito, B.; Drame, M.; Reveil, B.; Duret, V.; Jolly, D.; Pham, B.N.; Cohen, J.H. Alzheimer's disease is associated with low density of the long CR1 isoform. *Neurobiol. Aging* **2015**, *36*, e1765–e1766. [CrossRef]
47. McLure, C.A.; Williamson, J.F.; Stewart, B.J.; Keating, P.J.; Dawkins, R.L. Indels and imperfect duplication have driven the evolution of human Complement Receptor 1 (CR1) and CR1-like from their precursor CR1 alpha: Importance of functional sets. *Hum. Immunol.* **2005**, *66*, 258–273. [CrossRef]
48. Logar, C.M.; Chen, W.; Schmitt, H.; Yu, C.Y.; Birmingham, D.J. A human CR1-like transcript containing sequence for a binding protein for iC4 is expressed in hematopoietic and fetal lymphoid tissue. *Mol. Immunol.* **2004**, *40*, 831–840. [CrossRef] [PubMed]

Article

Set-Based Rare Variant Expression Quantitative Trait Loci in Blood and Brain from Alzheimer Disease Study Participants

Devanshi Patel [1,2], Xiaoling Zhang [2], John J. Farrell [2], Kathryn L. Lunetta [3] and Lindsay A. Farrer [1,2,3,4,5,6,*]

1. Bioinformatics Graduate Program, Boston University, Boston, MA 02215, USA; dpatel@bu.edu
2. Department of Medicine (Biomedical Genetics), Boston University School of Medicine, Boston, MA 02118, USA; zhangxl@bu.edu (X.Z.); farrell@bu.edu (J.J.F.)
3. Department of Biostatistics, Boston University School of Public Health, Boston, MA 02118, USA; klunetta@bu.edu
4. Department of Ophthalmology, Boston University School of Medicine, Boston, MA 02118, USA
5. Department of Neurology, Boston University School of Medicine, Boston, MA 02118, USA
6. Department of Epidemiology, Boston University School of Public Health, Boston, MA 02118, USA
* Correspondence: farrer@bu.edu

Abstract: Because studies of rare variant effects on gene expression have limited power, we investigated set-based methods to identify rare expression quantitative trait loci (eQTL) related to Alzheimer disease (AD). Gene-level and pathway-level cis rare-eQTL mapping was performed genome-wide using gene expression data derived from blood donated by 713 Alzheimer's Disease Neuroimaging Initiative participants and from brain tissues donated by 475 Religious Orders Study/Memory and Aging Project participants. The association of gene or pathway expression with a set of all cis potentially regulatory low-frequency and rare variants within 1 Mb of genes was evaluated using SKAT-O. A total of 65 genes expressed in the brain were significant targets for rare expression single nucleotide polymorphisms (eSNPs) among which 17% (11/65) included established AD genes *HLA-DRB1* and *HLA-DRB5*. In the blood, 307 genes were significant targets for rare eSNPs. In the blood and the brain, *GNMT, LDHC, RBPMS2, DUS2,* and *HP* were targets for significant eSNPs. Pathway enrichment analysis revealed significant pathways in the brain (n = 9) and blood (n = 16). Pathways for apoptosis signaling, cholecystokinin receptor (CCKR) signaling, and inflammation mediated by chemokine and cytokine signaling were common to both tissues. Significant rare eQTLs in inflammation pathways included five genes in the blood (*ALOX5AP, CXCR2, FPR2, GRB2, IFNAR1*) that were previously linked to AD. This study identified several significant gene- and pathway-level rare eQTLs, which further confirmed the importance of the immune system and inflammation in AD and highlighted the advantages of using a set-based eQTL approach for evaluating the effect of low-frequency and rare variants on gene expression.

Keywords: Alzheimer disease; expression quantitative trait loci (eQTL); rare variants; set-based eQTL; SKAT-O; pathways; immune system; inflammation; ROSMAP; ADNI

1. Introduction

Late-onset Alzheimer disease (AD) is the most common type of dementia that affects an estimated 5.7 million individuals aged 65 years and older in the United States, with the number projected to rise to 14 million by 2050 [1]. AD is highly heritable (h2 = 58–79%) [2], but common variants explain only one-third of the genetic portion of AD risk [2]. Highly penetrant rare variants may account for some of the missing heritability [3]. Whole-exome sequencing studies have identified robust AD associations with rare missense variants in *TREM2, AKAP9, UNC5C, ZNF655, IGHG3, CASP7* and *NOTCH3* [4–9], and it is expected that more AD-related rare variants will be identified by whole-genome sequencing (WGS) studies, because some rare variants, including those in non-coding regions, likely contribute to AD risk. However, identification of genes that are impacted by these rare variants, and thus likely have a functional role in AD, remains challenging.

Some AD risk variants are associated with gene expression, as demonstrated by recent expression quantitative trait locus (eQTL) studies [10,11]. Rare variants may contribute to extreme gene expression within a single tissue or across multiple tissues [12–15]. However, genome-wide studies of rare eQTLs are generally underpowered to obtain significant results. Although gene-based tests, which test the aggregate effects of multiple variants, are commonly used to evaluate the association of a disease with rare variants, only a few studies have applied this approach to the analysis of rare eQTLs. Several eQTL studies employed set-based approaches including testing gene expression with multiple single nucleotide polymorphisms (SNPs) chosen by variable selection [16,17] using a gene-based partial least-squares method to correlate multiple gene transcript probes with multiple SNPs [18], and identifying variants associated with transcript and protein modules [19]. These applications were not focused on rare variants, but still afforded higher power with a potential to find significant associations with low-frequency variants.

Few studies have applied a set-based eQTL method for rare variants. Recently, Lutz et al. applied burden and set-based (sequence) kernel association (SKAT) tests to normalize read counts in RNA-sequence (RNA-seq) studies [20]. In this study, we performed a gene-based cis-eQTL analysis using expression data derived from human blood and brain tissue to identify genes that contain a set of potentially regulatory low-frequency and rare variants (minor allele frequency (MAF) < 0.05) that are significantly associated with their expression. Although this design focused on rare variants, and thus has low power to detect expression differences between AD cases and controls, the set-based method can potentially discriminate AD-related targets among a group of genes located within 1 Mb from the expression single nucleotide polymorphisms (eSNPs) that were previously associated with the risk of AD. We also applied a pathway-based approach to determine which genes contribute most to the overall gene expression profile of a significant pathway containing a set of co-expressed functionally related genes.

2. Materials and Methods

2.1. Study Cohorts

The Alzheimer's Disease Neuroimaging Initiative (ADNI) is a multisite longitudinal study that began enrolling subjects in 2004, and includes persons with AD, mild cognitive impairment (MCI), and normal cognitive functioning [21]. Affymetrix Human Genome U219 array gene expression data derived from whole blood, whole-genome sequence (WGS) data, and phenotype data were downloaded from a public-access database (http://www.loni.usc.edu (accessed on 11 December 2018)). The portion of the sample included in this study included 207 AD cases, 284 MCI cases, 194 controls, and 28 individuals with missing dementia status.

The Religious Orders Study (ROS)/Memory and Aging Project (MAP) also contributed to this research. ROS enrolled older nuns and priests from across the US without known dementia for a longitudinal clinical analysis and brain donation. MAP enrolled older subjects without dementia from retirement homes, who agreed to brain donation at the time of death [22,23]. RNA-sequence data, including gene expression information derived from dorsolateral prefrontal cortex area tissue donated by 475 participants (281 autopsy-confirmed AD cases and 194 controls), as well as WGS data included in this study, were obtained from the AMP-AD knowledge portal (https://www.synapse.org/#!Synapse:syn3219045 (accessed on 1 July 2018)) [24]. Characteristics of subjects from both cohorts are provided in Table 1.

Table 1. Characteristics of subjects in the Religious Orders Study/Memory and Aging Project (ROSMAP) and Alzheimer's Disease Neuroimaging Initiative (ADNI) datasets.

Dataset	Race	N	AD Cases	MCI Cases	Controls	Female	Age *
ROSMAP (Brain)	NHW 98% AA 2% Other <0.01%	475	281	0	194	63%	85.9 (4.8)
ADNI (Blood)	NHW 93% AA 4% Other 3%	713	207	284	222	44%	76.3 (8.1)

NHW—non-Hispanic white, AA—African American. * mean (standard deviation).

2.2. Data Processing

ADNI microarray gene expression data were normalized and log-transformed using limma [25]. ROSMAP RNA-seq data were normalized and then log-transformed using a previously described pipeline [26]. The log-transformed expression data were evaluated using surrogate variable analysis (SVA) [27] to obtain surrogate variables for global technical effects and hidden effects, which were included as covariates in the analysis models for eQTL discovery. Additional filtering steps of GWAS and gene expression data included eliminating 167 ROSMAP and 96 ADNI subjects with missing data (resulting in the sample sizes reported in Table 1), restricting gene expression data to protein-coding genes (12,971 genes in ROSMAP and 16,025 genes in ADNI), and selecting only bi-allelic low-frequent and rare variants (MAF $\leq$ 0.05) with a variant call rate of >95%.

2.3. Functional Annotation of Variants

Variants in the ADNI and ROSMAP WGS datasets were annotated using CADD v1.6 [28] and GWAVA v1.0 software [29]. Combined Annotation-Dependent Depletion (CADD) scores prioritize functional, deleterious, and disease-causal coding and non-coding variants by integrating multiple annotations into one score by contrasting variants that survived natural selection with simulated mutations [28]. A scaled CADD score of 10 or greater indicates a raw score in the top 10% of all possible reference genome single nucleotide variants (SNVs), and a score of 20 or greater indicates a raw score in the top 1% [28]. Genome-Wide Annotation of Variants (GWAVA) scores predict the functional impact of non-coding genetic variants based on annotations of non-coding elements and genome-wide properties, such as evolutionary conservation and GC-content, in the range of 0–1 with mutations scored >0.5 identified as "functional" and those scored $\leq$0.5 as "non-functional" [29]. Genomic coordinates of variants in the ADNI dataset that were established using genome build GRCh38 were converted to build hg19 using liftOver software (https://genome.ucsc.edu/cgi-bin/hgLiftOver (accessed on 3 November 2018)). Both ADNI and ROSMAP WGS variants were matched by chromosome, position, reference, and alternate alleles. Variants having a CADD score >15 or a GWAVA region score >0.5 were annotated as having a potential regulatory function.

2.4. Set-Based eQTL Analysis

The sequence of steps to identify set-based eQTLs in the blood and brain is shown in Figure 1.

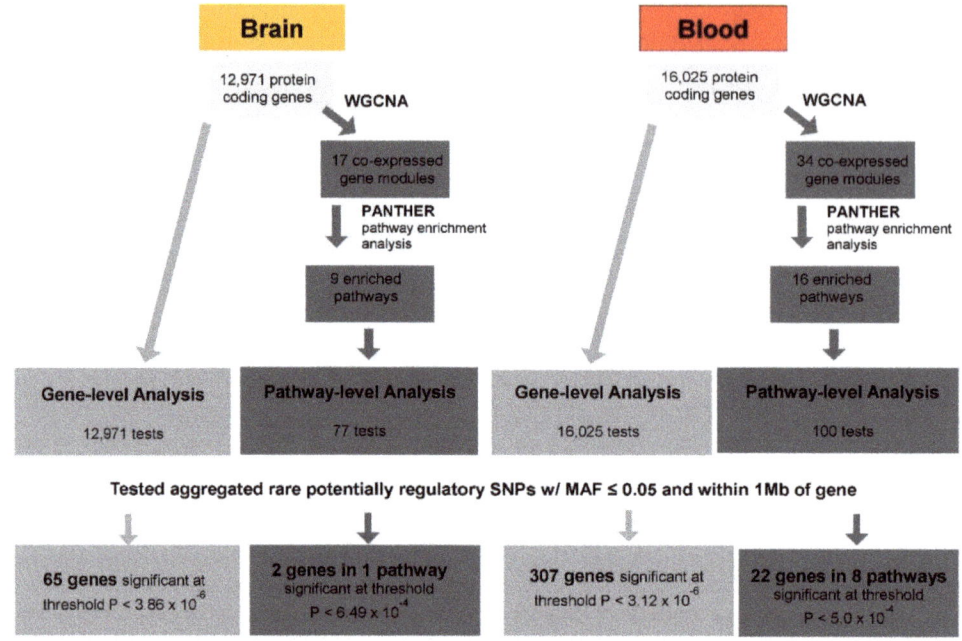

Figure 1. Overview of set-based rare expression quantitative trait loci (eQTL) analysis. Gene-level tests were performed for each protein-coding gene using an aggregate of all potentially regulatory single nucleotide polymorphisms (SNPs) with minor allele frequency ≤0.05 within 1 Mb of each gene. Pathway-level analysis was carried out in two steps. First, the weighted gene co-expression network analysis (WCGNA) method was applied to identify co-expressed gene modules. Next, pathway enrichment analysis was conducted using the Protein Analysis Through Evolutionary Relationships (PANTHER) tool to identify significantly enriched pathways in these gene modules, and pathway-level tests were then performed on each enriched pathway, including the aggregated SNPs for each gene in the module. Results were considered significant ($p < 0.05$) after applying a Bonferroni correction.

2.4.1. Gene-Level cis-eQTL Analysis

For common variants, eQTL analysis entails testing the association of expression of one gene with one variant. Gene-level eQTL analysis was performed by testing the association of expression of one gene with aggregated cis-regulatory variants, limited to those with a frequency of <0.05 and located in or within 1 Mb of the gene. Gene-based tests were performed using the SKAT-O method, which combines the variance component (SKAT) approach and burden tests into one test with optimal power [30]. We implemented SKAT-O tests for set-based eQTL analysis by considering the gene expression value as the outcome, with the aggregated rare variant count as the predictor. The regression model for analyses of the ROSMAP data also included covariates for age, sex, post-mortem interval (PMI), study (ROS or MAP), and a term for a surrogate variable (SV1), derived from the gene expression data matrix to account for unmeasured/hidden technical effects on gene expression using surrogate variable analysis (SVA) [27]. Model covariates for analyses of the ADNI data included baseline age, sex, RNA integrity number (RIN), year of blood sample collection, and SV1. SKAT-O was implemented with group-wise tests using EPACTS software (https://genome.sph.umich.edu/wiki/EPACTS (accessed on 9 March 2021)) with the following parameter specifications: epacts group—vcf [specific chr genome vcf.gz file]\—groupf [file of aggregated rare variants]—out [out file]\—ped [gene expression file]—max-maf 0.05\—pheno $gene—cov Age_baseline—cov Sex—cov RIN—cov Year of Collection—cov SV1—test skat—skat-o—run 8. The significance threshold after adjusting for the number of genes tested was 3.86×10^{-6} (0.05/12,971) for analyses of the ROSMAP data and 3.12×10^{-6} (0.05/16,024) for analyses of the ADNI data (Figure 1). To identify sentinel variants that contribute the majority of the evidence for significant gene-based results, eQTL tests were performed for all significant genes and each individual potentially regulatory rare variant (MAF $\leq$ 0.05) within 1Mb of the gene using linear regression models with the above covariates in R [31] for each cis-regulatory variant. The significance threshold after adjusting for the number of unique gene-SNP eQTLs was 1.83×10^{-6} (0.05/27,393) for analyses of the ROSMAP data and 1.17×10^{-7} (0.05/425,995) for analyses of the ADNI data.

2.4.2. Pathway-Level cis-eQTL Analysis

Pathway-level eQTL analysis was employed to test the association of a pathway, containing many genes, with sets of variants in each of the genes in the pathway one at a time. First, modules of co-expressed genes were identified using the Weighted Gene Co-expression Network Analysis (WGCNA) method implemented in R [32], including all protein-coding genes that were expressed in the ADNI and ROSMAP datasets. Analyses were conducted using the default parameters (soft-threshold power β = 6.00, deepSplit = 2 (medium sensitivity), a minimum module size of 20, and a merge cut height of 0.15) that were recommended by the developers of the software [32] and applied in another AD study [33]. Each gene module can be summarized quantitatively by a module eigengene (ME) value derived from principal component analysis. The ME is considered to be representative of gene expression profiles in a gene module. Next, gene-set pathway enrichment analysis was performed using the Protein Analysis Through Evolutionary Relationships (PANTHER) software tool [34] to determine which pathways were significantly enriched in the gene modules identified from the WGCNA for pathway-level eQTL analysis. Significance of the enriched pathways was determined by the Fisher's Exact test with a false discovery rate (FDR) of <0.05. Pathway-level eQTL analysis was performed for each significantly enriched pathway. The association of the ME value and each gene in the module was tested individually using all potentially regulatory rare cis-SNPs (MAF < 0.05). Models included the same covariates and parameter specifications as described for the gene-level eQTL tests and were analyzed using the SKAT-O method implemented in EPACTS. A total of 77 genes in 9 enriched pathways were evaluated in the ROSMAP dataset, and 100 genes in 16 enriched pathways were evaluated in the ADNI dataset. After correction for the number of genes that were tested, the thresholds for significant pathway-level rare eQTLs were $p < 6.49 \times 10^{-4}$ in the ROSMAP dataset and $p < 5.0 \times 10^{-4}$ in the ADNI dataset (Figure 1).

2.4.3. Comparison of Rare and Common eQTLs

To determine whether both common variants and gene-level aggregated rare/low-frequency variants target expression of the same genes, we evaluated the overlap in significant gene-based cis-eQTLs with those involving common variants (MAF > 0.05) within 1 Mb of protein-coding genes that were obtained previously from the Framingham Heart Study (blood) and ROSMAP (brain) gene expression datasets [26]. These comparisons not only indicated which eGenes are regulated by rare and/or common variants, but also determined whether multiple variants can separately up- or down-regulate expression of the same gene.

3. Results

3.1. Gene-Level eQTL Associations

In the gene-level eQTL analysis, aggregating on average 416 unique low-frequency and rare variants for each gene, 65 significant gene-level eQTLs ($p < 3.86 \times 10^{-6}$) were identified in the brain (Figure 1, Table S1). Eight of these genes, including established AD genes *HLA-DRB1* [35] and *HLA-DRB5* [36], are located in or near the major histocompatibility locus. By comparison, 307 significant gene-level eQTLs, with an average of 678 unique variants, were observed in blood at $p < 3.12 \times 10^{-6}$ (Figure 1, Table S2). Among these genes,

ABCA7, *ECHDC3*, and *MS4A6A* are known AD loci [35,36]. The genes *GNMT*, *LDHC*, *RBPMS2*, *DUS2*, and *HP* were significant in both the brain and blood (Table 2), noting that the evidence for *RBPMS2* was stronger in the blood ($p = 1.69 \times 10^{-36}$) than the brain ($p = 9.90 \times 10^{-8}$).

Table 2. Significant gene-level eQTLs common to blood and brain.

Chr	Begin Position	End Position	Gene	Brain			Blood		
				CVar +	Unique Var ^	*p*-Value	CVar +	Unique Var ^	*p*-Value
6	41,942,338	43,929,364	GNMT	671	437	1.85×10^{-6}	1006	640	2.87×10^{-7}
11	17,434,230	19,468,040	LDHC	429	273	2.07×10^{-7}	762	473	2.25×10^{-10}
15	64,039,999	66,063,761	RBPMS2	404	249	9.90×10^{-8}	648	417	1.69×10^{-36}
16	67,034,867	69,106,452	DUS2	714	482	1.98×10^{-6}	1085	723	6.41×10^{-08}
16	71,090,452	73,094,829	HP	741	461	2.28×10^{-9}	1206	750	2.43×10^{-11}

+ Cumulative number of variants. ^ Number of unique variants. Chromosome and map position according to GRCh37 assembly.

3.2. Variant-Level eQTL Associations

Examination of the variant-level eQTL associations for the 65 significant genes in the brain identified 61 significant eGene-eSNP eQTL pairs, involving 22 unique eGenes (Table S3). By a very wide margin, the most significant eQTL pair featured rs772849040 located in *NFAT5*, which targeted *DDX19A-DDX19B* ($p \leq 1.0 \times 10^{-314}$). *DDX19A-DDX19B* was also a significant eGene for rs17881635 located in COG4 ($p = 6.26 \times 10^{-23}$). *COPZ1* and *TMPRSS6* were both significant eGenes for seven eSNPs each. A much larger number of eQTL pairs ($n = 832$) were significant in the blood, in which 185 eGenes were unique (Table S4). Four of these genes had 20 or more significant eSNPs: *KRT79* ($n = 36$), *TAC3* ($n = 32$), *CDK12* ($n = 24$), and *SOS1* ($n = 20$). *LDHC* was a significant eGene for two eSNPs in the blood (rs117652970, $p = 1.12 \times 10^{-21}$ and rs17579565, $p = 8.26 \times 10^{-21}$) and a third eSNP in the brain rs773835421, $p = 1.60 \times 10^{-6}$). Adjacent genes *DHRS4* and its homolog *DHRS4L2* were significant eGene targets for 17 eSNPs. Similarly, three SNPs were each significant eQTLs paired with *ATP6V0D1* and *CMTM2*, and four SNPs were each significant eQTLs paired with *IKZF3* and *GSDMA*. In the brain, rs1260874991 and rs1405001784 were significant eSNPs for two zinc finger protein genes (*ZNF101* and *ZNF103*).

3.3. Pathways Enriched in the Brain and Blood

Pathway enrichment analysis of each gene module revealed 9 significant enriched pathways in the brain and 16 in the blood (Table 3). The apoptosis signaling, cholecystokinin receptor (CCKR) signaling map, and inflammation mediated by chemokine and cytokine signaling pathways were enriched in both the brain and blood. Focusing on genes in the significantly enriched pathways in the brain, the aggregated rare variants in *CCL7* and *CCL8* were associated with the inflammation mediated by chemokine and cytokine signaling pathway ($p = 1.84 \times 10^{-5}$ and $p = 4.50 \times 10^{-4}$, respectively, Table 4). In total, 6 of the 22 genes that contained significant aggregated rare eQTLs associated with pathway expression in the blood were members of the same inflammation pathway: *ALOX5AP* ($p = 1.26 \times 10^{-4}$), *CXCR2* ($p = 1.53 \times 10^{-6}$), *FPR2* ($p = 1.25 \times 10^{-4}$), *GRB2* ($p = 6.04 \times 10^{-7}$), *IFNAR1* ($p = 1.98 \times 10^{-5}$), and *RAF1* ($p = 2.11 \times 10^{-5}$) (Table 4). Furthermore, *CFLAR* ($p = 2.42 \times 10^{-4}$), *TMBIM6* ($p = 4.48 \times 10^{-4}$), and *TNFRSF10C* ($p = 8.77 \times 10^{-5}$) were significant rare variant eQTLs in apoptosis signaling pathways in the blood. Significant aggregated rare variant eQTLs were observed with *ALOX5AP* in both gene-level ($p = 2.20 \times 10^{-10}$) and pathway-level ($p = 1.26 \times 10^{-4}$) analyses (Table 4).

Table 3. Significant pathway enrichment in gene modules in the brain and blood.

Pathway	# Genes in Pathway	Gene Module	# Module Genes in Pathway	Module Genes Expected # of Genes *	Module Genes Fold Enrichment †	+/−	Uncorrected p-Value	FDR
BRAIN								
Apoptosis signaling	77	7	12	1.64	7.3	+	3.09×10^{-7}	5.01×10^{-5}
Toll receptor signaling	32	8	6	0.46	12.97	+	1.45×10^{-5}	2.36×10^{-3}
Wnt signaling	235	4	21	7.35	2.86	+	3.49×10^{-5}	5.65×10^{-3}
Cadherin signaling	127	4	14	3.97	3.53	+	9.59×10^{-5}	7.77×10^{-3}
CCKR signaling map	111	7	10	2.37	4.22	+	2.22×10^{-4}	1.20×10^{-2}
Gonadotropin-releasing hormone receptor	152	4	14	4.75	2.95	+	5.28×10^{-4}	2.14×10^{-2}
p53	62	7	7	1.32	5.29	+	5.76×10^{-4}	2.33×10^{-2}
Inflammation mediated by chemokine and cytokine signaling	173	16	5	0.51	9.89	+	1.54×10^{-4}	2.50×10^{-2}
Angiogenesis	126	4	12	3.94	3.05	+	9.95×10^{-4}	3.22×10^{-2}
BLOOD								
Blood coagulation	43	24	8	0.27	29.9	+	8.22×10^{-10}	1.34×10^{-7}
Parkinson's disease	85	15	7	0.79	8.82	+	2.28×10^{-5}	3.72×10^{-3}
Inflammation mediated by chemokine and cytokine signaling	237	14	11	2.27	4.84	+	2.50×10^{-5}	4.08×10^{-3}
T-cell activation	73	32	4	0.18	22.02	+	3.87×10^{-5}	6.30×10^{-3}
B-cell activation	66	12	6	0.66	9.08	+	7.89×10^{-5}	1.29×10^{-2}
PDGF signaling	127	12	7	1.27	5.5	+	3.79×10^{-4}	2.06×10^{-2}
Apoptosis signaling	112	5	12	3.15	3.81	+	1.51×10^{-4}	2.47×10^{-2}
JAK/STAT signaling	17	7	4	0.28	14.33	+	3.21×10^{-4}	2.62×10^{-2}
Ras	64	5	8	1.8	4.44	+	7.61×10^{-4}	3.10×10^{-2}
CCKR signaling map	164	5	14	4.61	3.04	+	3.94×10^{-4}	3.21×10^{-2}
Angiotensin II-stimulated signaling through G proteins and β-arrestin	33	5	6	0.93	6.47	+	6.14×10^{-4}	3.34×10^{-2}
Histamine H2 receptor-mediated signaling	24	5	5	0.67	7.41	+	1.03×10^{-3}	3.37×10^{-2}
Inflammation mediated by chemokine and cytokine signaling	237	24	7	1.47	4.75	+	7.94×10^{-4}	4.32×10^{-2}
Heme biosynthesis	11	6	4	0.25	15.93	+	2.74×10^{-4}	4.47×10^{-2}
Integrin signaling	180	7	11	2.96	3.72	+	2.84×10^{-4}	4.62×10^{-2}
Inflammation mediated by chemokine and cytokine signaling	237	20	8	1.78	4.48	+	5.07×10^{-4}	8.26×10^{-2}

* Number of genes expected in the gene module by chance based on the total set of genes in the pathway as determined for the ROSMAP and ADNI datasets. † fold enrichment = # module genes in pathway/expected number of genes in module.

Table 4. Significant pathway-level eQTLs in the brain or blood by aggregating cis rare variants.

CHR	Begin Position	End Position	Gene	CVAR +	Unique VAR ^	p-Value	Gene Module	Pathway
17	31,600,172	33,592,552	CCL7 *	340	206	1.84×10^{-5}	16	Inflammation mediated by chemokine and cytokine signaling
17	31,648,819	33,621,655	CCL8 *	319	195	4.50×10^{-4}	16	Inflammation mediated by chemokine and cytokine signaling
17	72,322,351	74,401,630	GRB2	1108	717	6.04×10^{-7}	14	Inflammation mediated by chemokine and cytokine signaling
2	217,992,496	220,001,949	CXCR2	943	564	1.53×10^{-6}	14	Inflammation mediated by chemokine and cytokine signaling
5	174,085,268	176,108,976	HRH2	335	196	9.07×10^{-6}	5	Histamine H2 receptor mediated signaling
1	25,859,096	27,901,441	RPS6KA1	1208	790	1.01×10^{-5}	5	Ras Pathway, CCKR signaling map
11	76,033,278	78,180,311	PAK1	565	355	1.83×10^{-5}	5	Ras Pathway, CCKR Signaling map
21	33,696,834	35,718,581	IFNAR1	525	332	1.98×10^{-5}	14	Inflammation mediated by chemokine and cytokine signaling
3	11,628,812	13,702,170	RAF1	431	255	2.11×10^{-5}	14	Inflammation mediated by chemokine and cytokine signaling
1	83,964,144	85,961,982	GNG5	589	336	3.43×10^{-5}	5	Histamine H2 receptor mediated signaling
9	115,150,150	117,160,754	ALAD	620	425	4.97×10^{-5}	6	Heme biosynthesis
1	44,478,672	46,476,606	UROD	1061	649	5.92×10^{-5}	6	Heme biosynthesis
19	13,202,507	15,228,794	PRKACA	767	501	7.60×10^{-5}	5	Histamine H2 receptor mediated signaling, CCKR signaling map
9	127,005,465	128998618	HSPA5	1235	692	7.91×10^{-5}	15	Parkinson disease
8	21,946,761	23968794	TNFRSF10C	817	520	8.77×10^{-5}	5	Apoptosis signaling
19	51,273,985	53272173	FPR2	440	280	1.23×10^{-4}	14	Inflammation mediated by chemokine and cytokine signaling
13	30,317,837	32,332,540	ALOX5AP	297	197	1.26×10^{-4}	14	Inflammation mediated by chemokine and cytokine signaling
2	200,984,212	203,030,077	CFLAR	637	404	2.42×10^{-4}	5	Apoptosis signaling
17	39,458,200	41,463,831	STAT5A	1093	708	3.08×10^{-4}	12	PDGF signaling
14	50,190,597	52,294,891	NIN	516	323	3.65×10^{-4}	12	PDGF signaling
12	49,108,257	51,158,233	TMBIM6	1082	716	4.48×10^{-4}	5	Apoptosis signaling

+ Cumulative number of variants; ^ Number of unique variants; * Results from brain (otherwise results from blood); Chromosome and map position according to GRCh37 assembly.

3.4. Gene Targets of eQTLs in the Brain and Blood

Comparison of significant rare and common eQTLs in each tissue (Figure 2) revealed 203 genes in the blood and 40 genes in the brain that were targets of rare and common eSNPs (Table S5), including 19 in the blood and 9 in the brain that have both been previously implicated in AD (Table 5). Three genes (*LDHC*, *RBPMS2*, and *HP*) are targets that were observed in significant rare and common eQTLs in the brain and the blood.

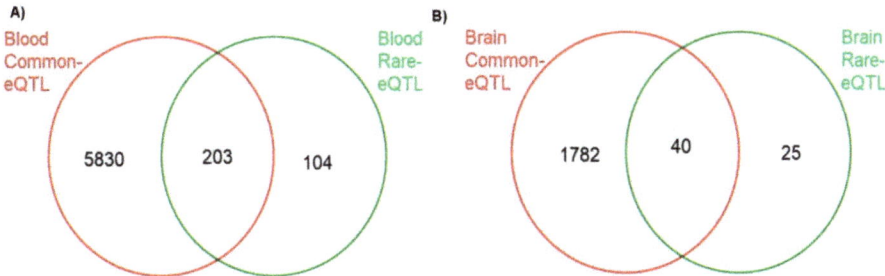

Figure 2. Overlap of significant genes in rare gene-level and common eQTLs in (**A**) the blood and (**B**) the brain.

Table 5. Genes previously implicated in AD whose expression is influenced by both rare and common SNPs.

eQTLs in Blood		eQTLs in Brain	
eGene	Reference	eGene	Reference
ABCA7 *	[36]	ACOT1	[37]
ADAMTSL4	[38]	HLA-A	[39]
ARRB2	[40]	HLA-DOB *	[26]
ATG7	[41]	HLA-DRB1 *	[35,41]
CD36	[42]	HLA-DRB5 *	[36]
CREB5	[43]	HP	[44]
CTNNAL1	[45]	POMC	[46]
ECHDC3 *	[35]	RNF39	[47,48]
HP	[44]	ZNF253	[49]
KF1B	[50,51]		
LRRC2	[52]		
MS4A6A *	[36]		
PADI2	[53]		
PDLIM5	[54]		
S100A12	[55]		
SPPL3	[56]		
TMEM51	[57]		
TREML4	[58]		
UBE4B	[59]		

* AD locus established by GWAS.

4. Discussion

Our study demonstrates that low-frequency and rare variants have a significant impact on both the expression of genes considered individually and the co-expression of genes in pathways. Our study highlights the value of the set-based rare-eQTL method because, similar to gene-based association tests, many novel significant genes we identified were not detected by the analysis of rare variants individually, which requires a much larger sample size. In addition, many of the most significant rare-variant findings involved genes with prior connections to AD through case-control comparisons using GWAS, gene expression, and functional studies.

Several of the most significant gene-level eQTL findings in the blood have previously been implicated in AD. *MS4A6A* ($p = 1.77 \times 10^{-22}$) is among a family of genes containing many SNPs that are associated with AD risk at the genome-wide level [35,36]). A meta-analysis of gene expression studies found that *NUMA1* ($p = 6.01 \times 10^{-76}$) was significantly upregulated in the hippocampus of AD cases [60], and another study showed that downregulation of *GAD1* ($p = 1.49 \times 10^{-58}$) was associated with reduced neuronal activity [61]. Follistatin, encoded by *FST* ($p = 4.02 \times 10^{-30}$), is a gonadal protein that inhibits the follicle-stimulating protein. The transmembrane protein, tomoregulin-2, contains follistatin-like modules and is found extensively in amyloid plaques in AD brains [62]. *KIF1B* ($p = 4.49 \times 10^{-21}$) expression is significantly increased in AD and is associated with accelerated progression in neurodegenerative diseases [50,51]. The established AD gene *ADAM10* [35] is downregulated by *SFRP1* ($p = 2.16 \times 10^{-20}$), which is significantly increased in the brain and cerebrospinal fluid (CSF) of AD patients [63]. *EXOC2* ($p = 6.19 \times 10^{-9}$) was identified as an AD age-of-onset modifier [64] and contains a rare missense variant that was observed in seven AD cases in an AD whole-exome sequencing study [9].

Four of the five significant gene-level rare eQTLs in the brain and blood (Table 1) have also been implicated in AD. *GNMT* expression has been detected in the hippocampus and its deficiency results in reduced neurogenic capacity, spatial learning, and memory impairment [65]. *LDHC* has differentially methylated regions in the blood in AD cases [66]. The overexpression of *DUS2* reduces $A\beta_{42}$ toxicity [67]. The acute-phase protein haptoglobin, encoded by *HP*, is significantly elevated among AD patients compared to healthy controls in serum [44,68] and CSF [69] in Asians and persons of European ancestry. The *HP* 1/1 genotype was associated with poorer cognitive function and greater cognitive decline than other *HP* genotypes in a sample of 466 African Americans with type 2 diabetes [70]. The RNA-binding protein *RBPMS2* has not been linked to AD but is a constituent of a leukocyte signature for traumatic brain injury [71].

We identified several pathways that are significantly enriched with genes involved in the CCKR signaling map, apoptosis signaling, and inflammation mediated by chemokine and cytokine signaling pathways, all of which have been linked to AD [72–74]. Wnt signaling, one of the significant pathways we observed in brain, suppresses tau phosphorylation and Aβ production/aggregation, inhibits *BACE1* expression, and promotes neuronal survival [75]. *HSPA5* ($p = 7.91 \times 10^{-5}$), one of the significant pathway-level eQTL findings, is involved in both amyloid precursor protein metabolism and neuronal death in AD [76].

Our rare-eQTL gene-level and pathway-level results confirm the substantial immune and inflammatory component to AD. Significant gene-level rare eQTLs in the brain included several HLA region loci linked to AD by GWAS (*HLA-DRB1* and *HLA-DRB5* [35,36]) and cell-type specific eQTL analysis (*HLA-DOB* [26]). *IL27* ($p = 1.69 \times 10^{-30}$) is a cytokine, and *CARD17* ($p = 6.73 \times 10^{-13}$) encodes a regulatory protein of inflammasomes, which are responsible for the activation of inflammatory responses [77]. Overall, 8 of the 21 significant pathway-level rare eQTLs involved genes which have roles in the inflammation mediated by the chemokine and cytokine signaling pathway. Chemokine levels were found to be significantly increased in serum, CSF, and brain tissue from AD cases [78]. Chemokine receptor *CXCR2* induces Aβ peptides [79]. Another gene in this group, *IFNAR1*, encodes the interferon α and β receptor subunit 1. Primary microglia isolated from the brains of *APP/PS1* mutant mice with ablated type-I interferon signaling have shown reduced levels of $A\beta_{1-42}$ [80]. In addition to being a significant pathway-level rare eQTL, *FPR2* is also very significant eQTL in the blood ($p = 1.22 \times 10^{-240}$), and more specifically, in interferon and anti-bacterial cells ($p = 3.81 \times 10^{-17}$) [26]. It is involved in the uptake and clearance of Aβ and contributes to innate immunity and inflammation [81]. *ALOX5AP* (a.k.a. *FLAP*) is expressed in microglia and encodes a protein which, with 5-lipoxygenase, is required for leukotriene synthesis. Leukotrienes are arachidonic acid metabolites which have been implicated in neuroinflammatory and amyloidogenesis processes in AD [82]. Pharmacological inhibition of FLAP in Tg2576 mice significantly reduced tau phosphorylation at multiple sites and increased post-synaptic density protein-95 and microtubule-associated protein 2 [83]. Growth factor receptor-bound protein 2, encoded by *GRB2*, is an adaptor protein that is involved in the trafficking of Aβ [84]. Although the inflammation pathway was implicated in the eQTL analysis in both the brain and blood, our results showed that the genes significantly contributing to pathway expression differed between the tissues. This suggests that AD-related inflammatory processes may differ in the blood and brain.

We observed significant eQTLs involving 27 target genes, previously implicated in AD through genetic and experimental approaches, which were paired with rare variants identified in this study and previously reported common variants [26] (Table 5). *HP* was the only gene in this group whose expression was influenced by rare and common eSNPs in both the blood and brain, and thus, it has notable potential as a blood-based biomarker reflecting AD-related gene expression changes in brain.

Although the set-based rare-eQTL method employed in this study has multiple strengths in comparison to the analysis of individual rare eQTLs (e.g., higher power, reduced multiple testing burden, and ability to detect the effects of variants with lower frequency), our results should be interpreted cautiously in light of several limitations. Comparisons between the brain and blood were not conducted using data from the same subjects, and thus may underestimate similarities across tissues. Also, brain expression patterns may reflect post-mortem changes unrelated to disease or cell-type specific expression [85]. The set-based method using SKAT-O allows for opposite effect directions of the constituent SNPs in the test; however, closer scrutiny of the individual SNPs is necessary to draw conclusions about the collective influence of rare variants on expression, as well as consistency of the effect direction across tissues. Our results, which were generated from analyses at the tissue level, do not account for patterns that are cell-type specific within the blood and brain, as we recently demonstrated for common individual variant eQTLs in these datasets [26]. In addition, it is unclear whether the set-based eQTL method applied in this study would behave similarly for rare (MAF < 0.01) and low-frequency (0.01 < MAF < 0.05) variants analyzed separately. Finally, although this investigation was conducted using tissue obtained from participants enrolled in studies of AD, the direct testing of the relevance of findings from the set-based tests of rare variants to AD status was not feasible, because the sample size was insufficient to have representation of the sentinel variants in both the case and control groups. This limitation is analogous to the difficulty encountered in the replication of the aggregated rare variant test findings in AD genetic association studies [7,8]. Thus, further studies of some genes are needed to establish their role in AD. Nonetheless, our study provided evidence favoring specific genes under previously established AD-association peaks whose expression may be differentially or concordantly regulated in the blood and brain (Table 5).

5. Conclusions

This study of gene-based and pathway-level rare eQTLs implicated novel genes that may have important roles in AD, found additional evidence supporting the contribution of immune/inflammatory pathways in AD, and demonstrated the utility of a set-based eQTL approach for assessing the role of rare variants in molecular mechanisms underlying the disease. The relevance of these findings to AD should be validated in larger samples with sufficient power for comparing patterns between AD cases and controls, as well as with functional experiments.

Supplementary Materials: The following are available online at https://www.mdpi.com/2073-4425/12/3/419/s1, Table S1: Gene-level rare cis-eQTLs in the brain ($p < 3.86 \times 10^{-6}$), Table S2: Individual SNP eQTLs in the brain ($p < 1.83 \times 10^{-6}$), Table S3: Gene-level rare cis-eQTLs in the blood ($p < 3.12 \times 10^{-6}$), Table S4: Individual SNP eQTLs in the blood ($p < 1.17 \times 10^{-7}$), Table S5: eGene targets of both rare and common eQTLs in the blood and brain.

Author Contributions: D.P. analyzed the data and prepared the figures and tables. J.J.F. provided database management and bioinformatics support. X.Z. and K.L.L. provided expert advice on the design and execution of the statistical genetic analyses. D.P. and L.A.F. wrote the manuscript. L.A.F. obtained the funding for this study. X.Z. and L.A.F. supervised the project. All authors have read and agreed to the published version of the manuscript.

Funding: This study was supported by NIH grants RF1-AG057519, 2R01-AG048927 U01-AG058654, P30-AG13846, 3U01-AG032984, U01-AG062602, and U19-AG068753. Collection of study data provided by the Rush Alzheimer's Disease Center, Rush University Medical Center, Chicago was supported through funding by NIA grants P30AG10161, R01AG15819, R01AG17917, R01AG30146, R01AG36836, U01AG32984, U01AG46152, U01AG61358, a grant from the Illinois Department of Public Health, and the Translational Genomics Research Institute. Collection and sharing of ADNI data were funded by the Alzheimer's Disease Neuroimaging Initiative (ADNI) (National Institutes of Health Grant U01 AG024904) and DOD ADNI (Department of Defense award number W81 X WH-12-2-0012). ADNI is funded by the National Institute on Aging, the National Institute of Biomedical Imaging and Bioengineering, and through generous contributions from the following: Alzheimer's Association; Alzheimer's Drug Discovery Foundation; Araclon Biotech; BioClinica, Inc.; Biogen Idec Inc.; Bristol-Myers Squibb Company; Eisai Inc.; Elan Pharmaceuticals, Inc.; Eli Lilly and Company; EuroImmun; F. Hoffmann-La Roche Ltd. and its affiliated company Genentech, Inc.; Fujirebio; GE Healthcare; IXICO Ltd.; Janssen Alzheimer's Immunotherapy Research & Development, LLC.; Johnson & Johnson Pharmaceutical Research & Development LLC.; Medpace, Inc.; Merck & Co., Inc.; Meso Scale Diagnostics, LLC.; NeuroR x Research; Neurotrack Technologies; Novartis Pharmaceuticals Corporation; Pfizer Inc.; Piramal Imaging; Servier; Synarc Inc.; and Takeda Pharmaceutical Company. The Canadian Institute of Health Research is providing funds to support ADNI clinical sites in Canada. Private sector contributions are facilitated by the Foundation for the National Institutes of Health (www.fnih.org (accessed on 9 March 2021)). The grantee organization is the Northern Alzheimer's Disease Cooperative Study at the University of Southern California, and the study is coordinated by the Alzheimer's Therapeutic Research Institute at the University of Southern California. ADNI data are disseminated by the Laboratory for Neuroimaging at the University of Southern California.

Institutional Review Board Statement: This study was approved by the Boston University Institutional Review Board.

Informed Consent Statement: Not applicable.

Data Availability Statement: The results published here are in whole or in part based on data obtained from the AD Knowledge Portal (https://adknowledgeportal.synapse.org (accessed on 1 July 2018)). ROSMAP study data were provided by the Rush Alzheimer's Disease Center, Rush University Medical Center, Chicago. ADNI data can be obtained by study investigators (adni.loni.usc.edu (accessed on 9 March 2021)).

Acknowledgments: Data used in preparation of this article were obtained from the Alzheimer's Disease Neuroimaging Initiative (ADNI) database (adni.loni.usc.edu). As such, the investigators within the ADNI contributed to the design and implementation of ADNI and/or provided data but did not participate in the analysis or writing of this report. A complete listing of ADNI investigators can be found at: http://adni.loni.usc.edu/wp-content/uploads/how_to_apply/ADNI_Acknowledgement_List.pdf (accessed on 9 March 2021).

Conflicts of Interest: The authors declare no conflict of interest.

Abbreviations

Aβ	amyloid-β
AD	Alzheimer disease
ADNI	Alzheimer's Disease Neuroimaging Initiative
eQTL	Expression quantitative trait locus
CADD	Combined annotation-dependent depletion
GWAVA	Genome-wide annotation of variants

MAF	Minor allele frequency
MCI	Mild cognitive impairment
ME	module eigengene
PANTHER	Protein Analysis Through Evolutionary Relationships
ROSMAP	Religious Orders Study/Memory and Aging Project
SNP	Single nucleotide polymorphism
SVA	Surrogate variable analysis
WCGNA	Weighted gene co-expression network analysis
WGS	Whole-genome sequence

References

1. Alzheimer's Association. Facts and Figures. 2020. Available online: https://alz.org/alzheimers-dementia/facts-figures (accessed on 20 June 2020).
2. Gatz, M.; Reynolds, C.A.; Fratiglioni, L.; Johansson, B.; Mortimer, J.A.; Berg, S.; Fisk, A.; Pedersen, N.L. Role of genes and environments for explaining Alzheimer disease. *Arch. Gen. Psychiatry* **2006**, *63*, 168–174. [CrossRef]
3. Sims, R.; van der Lee, S.; Naj, A.C.; Bellenguez, C.; Badarinarayan, N.; Jakobsdottir, J.; Kunkle, B.W.; Boland, A.; Raybould, R.; Bis, J.C.; et al. Rare coding variants in *PLCG2*, *ABI3*, and *TREM2* implicate microglial-mediated innate immunity in Alzheimer's disease. *Nat. Genet.* **2017**, *49*, 1373–1384. [CrossRef] [PubMed]
4. Guerreiro, R.; Wojtas, A.; Bras, J.; Carrasquillo, M.; Rogaeva, E.; Majounie, E.; Cruchaga, C.; Sassi, C.; Kauwe, J.S.K.; Younkin, S.; et al. TREM2 variants in Alzheimer's disease. *N. Engl. J. Med.* **2013**, *368*, 117–127. [CrossRef] [PubMed]
5. Logue, M.W.; Schu, M.; Vardarajan, B.N.; Farrell, J.; Bennett, D.A.; Buxbaum, J.D.; Byrd, G.S.; Ertekin-Taner, N.; Evans, D.; Foroud, T.; et al. Two rare AKAP9 variants are associated with Alzheimer disease in African Americans. *Alzheimers Dement.* **2014**, *10*, 609–618. [CrossRef]
6. Wetzel-Smith, M.K.; Hunkapiller, J.; Bhangale, T.R.; Srinivasan, K.; Maloney, J.A.; Atwal, J.K.; Sa, S.M.; Yaylaoglu, M.B.; Foreman, O.; Ortmann, W.; et al. A rare mutation in UNC5C predisposes to late-onset Alzheimer's disease and increases neuronal cell death. *Nat. Med.* **2014**, *20*, 1452–1457. [CrossRef] [PubMed]
7. Bis, J.; Jian, X.; Kunkle, B.; Chen, Y.; Hamilton-Nelson, K.; Bush, W.S.; Salerno, W.J.; Lancour, D.; Ma, Y.; Renton, A.E.; et al. Whole exome sequencing study identifies novel rare and common Alzheimer's-associated variants involved in immune response and transcriptional regulation. *Mol. Psychiatry* **2020**, *25*, 1901–1903. [CrossRef]
8. Zhang, X.; Zhu, C.; Beecham, G.; Vardarajan, B.N.; Ma, Y.; Lancour, D.; Farrell, J.J.; Chung, J. A rare missense variant of CASP7 is associated with familial late-onset Alzheimer's disease. *Alzheimers Dement.* **2020**, *15*, 441–452. [CrossRef]
9. Patel, D.; Mez, J.; Vardarajan, B.N.; Staley, L.; Chung, J.; Zhang, X.; Farrell, J.J.; Rynkiewicz, M.J.; Cannon-Albright, L.A.; Teerlink, C.C.; et al. Association of rare coding mutations with Alzheimer disease and other dementias among adults of European ancestry. *JAMA Netw. Open* **2019**, *2*, e191350. [CrossRef]
10. Rao, S.; Ghani, M.; Guo, Z.; Deming, Y.; Wang, K.; Sims, R.; Mao, C.; Yao, Y.; Cruchaga, C.; Stephan, D.A.; et al. An APOE-independent cis-eSNP on chromosome 19q13.32 influences tau levels and late-onset Alzheimer's disease risk. *Neurobiol. Aging* **2018**, *66*, 178.e1–178.e8. [CrossRef]
11. Zou, F.; Carrasquillo, M.M.; Pankratz, V.S.; Belbin, O.; Morgan, K.; Allen, M.; Wilcox, S.L.; Ma, L.; Walker, L.P.; Kouri, N.; et al. Gene expression levels as endophenotypes in genome-wide association studies of Alzheimer disease. *Neurology* **2010**, *74*, 480–486. [CrossRef]
12. Li, X.; Kim, Y.; Tsang, E.K.; Davis, J.R.; Damani, F.N.; Chiang, C.; Hess, G.T.; Zappala, Z.; Strober, B.J.; Scott, A.J.; et al. The impact of rare variation on gene expression across tissues. *Nature* **2017**, *550*, 239–243. [CrossRef]
13. Zhao, J.; Akinsanmi, I.; Arafat, D.; Cradick, T.J.; Lee, C.M.; Banskota, S.; Marigorta, U.M.; Bao, G.; Gibson, G. A burden of rare variants associated with extremes of gene expression in human peripheral blood. *Am. J. Hum. Genet.* **2016**, *98*, 299–309. [CrossRef]
14. Montgomery, S.B.; Lappalainen, T.; Gutierrez-Arcelus, M.; Dermitzakis, E.T. Rare and common regulatory variation in population-scale sequenced human genomes. *PLoS Genet.* **2011**, *7*, e1002144. [CrossRef]
15. Zeng, Y.; Wang, G.; Yang, E.; Ji, G.; Brinkmeyer-Langford, C.L.; Cai, J.J. Aberrant gene expression in humans. *PLoS Genet.* **2015**, *11*, e1004942. [CrossRef] [PubMed]
16. Daye, Z.J.; Chen, J.; Li, H. High-dimensional heteroscedastic regression with an application to eQTL data analysis. *Biometrics* **2012**, *68*, 316–326. [CrossRef] [PubMed]
17. Sun, W.; Ibrahim, J.G.; Zou, F. Genomewide multiple-loci mapping in experimental crosses by iterative adaptive penalized regression. *Genetics* **2010**, *185*, 349–359. [CrossRef] [PubMed]
18. Yang, H.; Lin, C.; Chen, C.; Chen, J.J. Applying genome-wide gene-based expression quantitative trait locus mapping to study population ancestry and pharmacogenetics. *BMC Genom.* **2014**, *15*, 319. [CrossRef] [PubMed]
19. Yang, M.Q.; Li, D.; Yang, W.; Zhang, Y.; Liu, J.; Tong, W. A gene module-based eQTL analysis prioritizing disease genes and pathways in kidney cancer. *Comput. Struct. Biotechnol. J.* **2017**, *15*, 463–470. [CrossRef] [PubMed]
20. Lutz, S.M.; Thwing, A.; Fingerlin, T. eQTL mapping of rare variant associations using RNA-seq data: An evaluation of approaches. *PLoS ONE* **2019**, *14*, e0223273. [CrossRef]

21. Alzheimer's Disease Neuroimaging Initiative. ADNI—Alzheimer's Disease Neuroimaging Initiative. Available online: http://adni.loni.usc.edu/ (accessed on 11 December 2018).
22. Bennett, D.A.; Schneider, J.A.; Arvanitakis, Z.; Wilson, R.S. Overview and findings from the religious orders study. *Curr. Alzheimer Res.* **2012**, *9*, 628–645. [CrossRef]
23. Bennett, D.A.; Schneider, J.A.; Buchman, A.S.; Barnes, L.L.; Wilson, R.S.; Boyle, P.A.; Wilson, R.S. Overview and findings from the Rush Memory and Aging Project. *Curr. Alzheimer Res.* **2012**, *9*, 646–663. [CrossRef]
24. AMP-AD Knowledge Portal. 2018. Available online: https://www.synapse.org/ (accessed on 1 July 2018).
25. Ritchie, M.E.; Phipson, B.; Wu, D.; Hu, Y.; Law, C.W.; Shi, W.; Smyth, G.K. limma powers differential expression analyses for RNA-sequencing and microarray studies. *Nucleic Acids Res.* **2015**, *43*, e47. [CrossRef]
26. Patel, D.; Zhang, X.; Farrell, J.; Chung, J.; Stein, T.D.; Lunetta, K.L.; Farrer, L.A. Cell-type specific expression quantitative trait loci associated with Alzheimer disease in blood and brain tissue. *MedRxiv* **2020**. [CrossRef]
27. Leek, J.T.; Johnson, W.E.; Parker, H.S.; Jaffe, A.E.; Storey, J.D. The sva package for removing batch effects and other unwanted variation in high-throughput experiments. *Bioinformatics* **2012**, *28*, 882–883. [CrossRef]
28. Rentzsch, P.; Witten, D.; Cooper, G.M.; Shendure, J.; Kircher, M. CADD: Predicting the deleteriousness of variants throughout the human genome. *Nucleic Acids Res.* **2019**, *47*, D886–D894. [CrossRef] [PubMed]
29. Ritchie, G.R.S.; Dunham, I.; Zeggini, E.; Flicek, P. Functional annotation of non-coding sequence variants. *Nat. Methods* **2014**, *11*, 294–296. [CrossRef] [PubMed]
30. Lee, S.; Emond, M.J.; Bamshad, M.J.; Barnes, K.C.; Rieder, M.J.; Nickerson, D.A.; Christiani, D.C.; Wurfel, M.H.; Lin, X.; NHLBI GO Exome Sequencing Project—ESP Lung Project Team; et al. Optimal unified approach for rare-variant association testing with application to small-sample case-control whole-exome sequencing studies. *Am. J. Hum. Genet.* **2012**, *91*, 224–237. [CrossRef]
31. R Core Team. R: A Language and Environment for Statistical Computing. R Foundation for Statistical Computing. Available online: http://www.R-project.org/ (accessed on 8 April 2020).
32. Langfelder, P.; Horvath, S. WGCNA: An R package for weighted correlation network analysis. *BMC Bioinform.* **2008**, *9*, 559. [CrossRef]
33. Zhang, B.; Gaiteri, C.; Bodea, L.; Wang, Z.; McElwee, J.; Podtelezhnikov, A.A.; Zhang, C.; Xie, T.; Tran, L.; Dobrin, R.; et al. Integrated systems approach identifies genetic nodes and networks in late-onset Alzheimer's disease. *Cell* **2013**, *153*, 707–720. [CrossRef]
34. Mi, H.; Huang, X.; Muruganujan, A.; Tang, H.; Mills, C.; Kang, D.; Thomas, P.D. PANTHER version 11: Expanded annotation data from Gene Ontology and Reactome pathways, and data analysis tool enhancements. *Nucleic Acids Res.* **2017**, *45*, D183–D189. [CrossRef]
35. Kunkle, B.W.; Grenier-Boley, B.; Sims, R.; Bis, J.C.; Damotte, V.; Naj, A.C.; Boland, A.; Vronskaya, M.; van der Lee, S.J.; Amlie-Wolf, A.; et al. Genetic meta-analysis of diagnosed Alzheimer's disease identifies new risk loci and implicates Aβ, tau, immunity and lipid processing. *Nat. Genet.* **2019**, *51*, 414–430. [CrossRef] [PubMed]
36. Lambert, J.C.; Ibrahim-Verbaas, C.A.; Harold, D.; Naj, A.C.; Sims, R.; Bellenguez, C.; DeStafano, A.L.; Bis, J.C.; Beecham, G.W.; Grenier-Boley, B.; et al. Meta-analysis of 74,046 individuals identifies 11 new susceptibility loci for Alzheimer's disease. *Nat. Genet.* **2013**, *45*, 1452–1458. [CrossRef]
37. Nagata, Y.; Hirayama, A.; Ikeda, S.; Shirahata, A.; Shoji, F.; Maruyama, T.; Kayano, M.; Bundo, M.; Hattori, K.; Yoshida, S.; et al. Comparative analysis of cerebrospinal fluid metabolites in Alzheimer's disease and idiopathic normal pressure hydrocephalus in a Japanese cohort. *Biomark. Res.* **2018**, *6*, 5. [CrossRef] [PubMed]
38. Hu, Y.; Xin, J.; Hu, Y.; Zhang, L.; Wang, J. Analyzing the genes related to Alzheimer's disease via a network and pathway-based approach. *Alzheimers Res. Ther.* **2017**, *9*, 29. [CrossRef] [PubMed]
39. Ma, S.L.; Tang, N.L.S.; Tam, C.W.C.; Lui, V.W.C.; Suen, E.W.C.; Chiu, H.F.K.; Lam, L.C.W. Association between HLA-A alleles and Alzheimer's disease in a southern Chinese community. *Dement. Geriatr. Cogn. Disord.* **2008**, *26*, 391–397. [CrossRef] [PubMed]
40. Jiang, T.; Yu, J.; Wang, Y.; Wang, H.; Zhang, W.; Hu, N.; Tan, L.; Sun, L.; Tan, M.-S.; Zhu, X.-C.; et al. The genetic variation of ARRB2 is associated with late-onset Alzheimer's disease in Han Chinese. *Curr. Alzheimer Res.* **2014**, *11*, 408–412. [CrossRef]
41. Uddin, M.S.; Stachowiak, A.; Mamun, A.A.; Tzvetkov, N.T.; Takeda, S.; Atanasov, A.G.; Bergantin, L.B.; Abdel-Daim, M.M.; Stankiewicz, A.M. Autophagy and Alzheimer's disease: From molecular mechanisms to therapeutic implications. *Front. Aging Neurosci.* **2018**, *10*, 4. [CrossRef]
42. Šerý, O.; Janoutová, J.; Ewerlingová, L.; Hálová, A.; Lochman, J.; Janout, V.; Khan, N.A.; Balcar, V.J. CD36 gene polymorphism is associated with Alzheimer's disease. *Biochimie* **2017**, *135*, 46–53. [CrossRef]
43. Nho, K.; Nudelman, K.; Allen, M.; Hodges, A.; Kim, S.; Risacher, S.L.; Apostolova, L.G.; Lin, K.; Lunnon, K.; Wang, X.; et al. Genome-wide transcriptome analysis identifies novel dysregulated genes implicated in Alzheimer's pathology. *Alzheimers Dement.* **2020**, *16*, 1213–1223. [CrossRef]
44. Song, I.; Kim, Y.; Chung, S.; Cho, H. Association between serum haptoglobin and the pathogenesis of Alzheimer's disease. *Intern. Med.* **2015**, *54*, 453–457. [CrossRef]
45. Hardingham, G.E.; Pruunsild, P.; Greenberg, M.E.; Bading, H. Lineage divergence of activity-driven transcription and evolution of cognitive ability. *Nat. Rev. Neurosci.* **2018**, *19*, 9–15. [CrossRef]
46. Shen, Y.; Tian, M.; Zheng, Y.; Gong, F.; Fu, A.K.Y.; Ip, N.Y. Stimulation of the hippocampal POMC/MC4R circuit alleviates synaptic plasticity impairment in an Alzheimer's disease model. *Cell Rep.* **2016**, *17*, 1819–1831. [CrossRef]

47. De Jager, P.L.; Srivastava, G.; Lunnon, K.; Burgess, J.; Schalkwyk, L.C.; Yu, L.; Eaton, M.L.; Keenan, B.T.; Ernst, J.; McCabe, C.; et al. Alzheimer's disease: Early alterations in brain DNA methylation at ANK1, BIN1, RHBDF2 and other loci. *Nat. Neurosci.* **2014**, *17*, 1156–1163. [CrossRef]
48. Smith, R.G.; Lunnon, K. DNA modifications and Alzheimer's disease. In *Neuroepigenomics in Aging and Disease*; Delgado-Morales, R., Ed.; Springer International Publishing: Cham, Switzerland, 2017; pp. 303–319.
49. Blalock, E.M.; Geddes, J.W.; Chen, K.C.; Porter, N.M.; Markesbery, W.R.; Landfield, P.W. Incipient Alzheimer's disease: Microarray correlation analyses reveal major transcriptional and tumor suppressor responses. *Proc. Natl. Acad. Sci. USA* **2004**, *101*, 2173–2178. [CrossRef] [PubMed]
50. Kreft, K.L.; van Meurs, M.; Wierenga-Wolf, A.F.; Melief, M.; van Strien, M.E.; Hol, E.M.; Oostra, B.A.; Laman, J.D.; Hintzen, R.Q. Abundant kif21b is associated with accelerated progression in neurodegenerative diseases. *Acta Neuropathol. Commun.* **2014**, *2*, 144. [CrossRef]
51. Hares, K.; Miners, J.S.; Cook, A.J.; Rice, C.; Scolding, N.; Love, S.; Wilkins, A. Overexpression of kinesin superfamily motor proteins in Alzheimer's disease. *J. Alzheimers Dis.* **2017**, *60*, 1511–1524. [CrossRef] [PubMed]
52. Raghavan, N.S.; Vardarajan, B.; Mayeux, R. Genomic variation in educational attainment modifies Alzheimer disease risk. *Neurol. Genet.* **2019**, *5*, e310. [CrossRef] [PubMed]
53. Arif, M.; Kato, T. Increased expression of PAD2 after repeated intracerebroventricular infusions of soluble Abeta(25-35) in the Alzheimer's disease model rat brain: Effect of memantine. *Cell Mol. Biol. Lett.* **2009**, *14*, 703–714. [CrossRef] [PubMed]
54. Herrick, S.; Evers, D.M.; Lee, J.; Udagawa, N.; Pak, D.T.S. Postsynaptic PDLIM5 / Enigma Homolog binds SPAR and causes dendritic spine shrinkage. *Mol. Cell Neurosci.* **2010**, *43*, 188. [CrossRef] [PubMed]
55. Shepherd, C.E.; Goyette, J.; Utter, V.; Rahimi, F.; Yang, Z.; Geczy, C.L.; Halliday, G.M. Inflammatory S100A9 and S100A12 proteins in Alzheimer's disease. *Neurobiol. Aging* **2006**, *27*, 1554–1563. [CrossRef] [PubMed]
56. El-Battari, A.; Mathieu, S.; Sigaud, R.; Prorok-Hamon, M.; Ouafik, L.; Jeanneau, C. Elucidating the roles of Alzheimer disease-associated proteases and the signal-peptide peptidase-like 3 (SPPL3) in the shedding of glycosyltransferases. *BioRxiv* **2018**, 317214.
57. Zhang, S.; Qin, C.; Cao, G.; Guo, L.; Feng, C.; Zhang, W. Genome-wide analysis of DNA methylation profiles in a senescence-accelerated mouse prone 8 brain using whole-genome bisulfite sequencing. *Bioinformatics* **2017**, *33*, 1591–1595. [CrossRef] [PubMed]
58. Gonzalez-Cotto, M.; Guo, L.; Karwan, M.; Sen, S.K.; Barb, J.; Collado, C.J.; Elloumi, F.; Palmieri, E.M.; Boelte, K.; Kolodgie, F.D.; et al. TREML4 promotes inflammatory programs in human and murine macrophages and alters atherosclerosis lesion composition in the apolipoprotein E deficient mouse. *Front. Immunol.* **2020**, *11*, 397. [CrossRef]
59. Gireud-Goss, M.; Reyes, S.; Tewari, R.; Patrizz, A.; Howe, M.D.; Kofler, J.; Waxham, M.N.; McCullough, L.D.; Bean, A.J. The ubiquitin ligase UBE4B regulates amyloid precursor protein ubiquitination, endosomal trafficking, and amyloid β42 generation and secretion. *Mol. Cell. Neurosci.* **2020**, *108*, 103542. [CrossRef]
60. Moradifard, S.; Hoseinbeyki, M.; Ganji, S.M.; Minuchehr, Z. Analysis of microRNA and Gene Expression Profiles in Alzheimer's Disease: A Meta-Analysis Approach. *Sci. Rep.* **2018**, *8*, 1–17. [CrossRef]
61. Gleichmann, M.; Zhang, Y.; Wood, W.H.; Becker, K.G.; Mughal, M.R.; Pazin, M.J.; van Praag, H.; Kobilo, T.; Zonderman, A.B.; Troncoso, J.C.; et al. Molecular changes in brain aging and Alzheimer's disease are mirrored in experimentally silenced cortical neuron networks. *Neurobiol. Aging* **2012**, *33*, 205.e1. [CrossRef] [PubMed]
62. Siegel, D.A.; Davies, P.; Dobrenis, K.; Huang, M. Tomoregulin-2 is found extensively in plaques in Alzheimer's disease brain. *J. Neurochem.* **2006**, *98*, 34–44. [CrossRef]
63. Esteve, P.; Rueda-Carrasco, J.; Mateo, M.I.; Martin-Bermejo, M.J.; Draffin, J.; Pereyra, G.; Sandonís, A.; Crespo, I.; Moreno, I.; Aso, E.; et al. Elevated levels of secreted-frizzled-related-protein 1 contribute to Alzheimer's disease pathogenesis. *Nat. Neurosci.* **2019**, *22*, 1258–1268. [CrossRef] [PubMed]
64. Vélez, J.I.; Lopera, F.; Sepulveda-Falla, D.; Patel, H.R.; Johar, A.S.; Chuah, A.; Tobón, C.; Rivera, D.; Villegas, A.; Cai, Y.; et al. APOE*E2 allele delays age of onset in PSEN1 E280A Alzheimer's disease. *Mol. Psychiatry* **2016**, *21*, 916–924. [CrossRef]
65. Carrasco, M.; Rabaneda, L.G.; Murillo-Carretero, M.; Ortega-Martínez, S.; Martínez-Chantar, M.L.; Woodhoo, A.; Luka, Z.; Wagner, C.; Lu, S.C.; Mato, J.M.; et al. Glycine N-methyltransferase expression in the hippocampus and its role in neurogenesis and cognitive performance. *Hippocampus* **2014**, *24*, 840–852. [CrossRef]
66. Lardenoije, R.; Roubroeks, J.A.Y.; Pishva, E.; Leber, M.; Wagner, H.; Iatrou, A.; Smith, A.R.; Smith, R.G.; Eijssen, L.M.T.; Kleineidam, L.; et al. Alzheimer's disease-associated (hydroxy)methylomic changes in the brain and blood. *Clin. Epigenetics* **2019**, *11*, 164. [CrossRef]
67. Chen, X.; Ji, B.; Hao, X.; Li, X.; Eisele, F.; Nyström, T.; Petranovic, D. FMN reduces Amyloid-β toxicity in yeast by regulating redox status and cellular metabolism. *Nat. Commun.* **2020**, *11*, 867. [CrossRef]
68. Zhu, C.; Jiang, G.; Chen, J.; Zhou, Z.; Cheng, Q. Serum haptoglobin in Chinese patients with Alzheimer's disease and mild cognitive impairment: A case-control study. *Brain Res. Bull.* **2018**, *137*, 301–305. [CrossRef]
69. Ayton, S.; Janelidze, S.; Roberts, B.; Palmqvist, S.; Kalinowski, P.; Diouf, I.; Belaidi, A.A.; Stomrud, E.; Bush, A.I.; Hansson, O. Acute phase markers in CSF reveal inflammatory changes in Alzheimer's disease that intersect with pathology, APOE ε4, sex and age. *Prog. Neurobiol.* **2021**, *198*, 101904. [CrossRef] [PubMed]

70. Méndez-Gómez, J.L.; Pelletier, A.; Rougier, M.; Korobelnik, J.; Schweitzer, C.; Delyfer, M.; Catheline, G.; Monfermé, S.; Dartigues, J.-F.; Delcourt, C.; et al. Association of retinal nerve fiber layer thickness with brain alterations in the visual and limbic networks in elderly adults without dementia. *JAMA Netw. Open* **2018**, *1*, e184406. [CrossRef]
71. Meng, Q.; Zhuang, Y.; Ying, Z.; Agrawal, R.; Yang, X.; Gomez-Pinilla, F. Traumatic brain injury induces genome-wide transcriptomic, methylomic, and network perturbations in brain and blood predicting neurological disorders. *EBioMedicine* **2017**, *16*, 184–194. [CrossRef] [PubMed]
72. Le Page, A.; Dupuis, G.; Frost, E.H.; Larbi, A.; Pawelec, G.; Witkowski, J.M.; Fulop, T. Role of the peripheral innate immune system in the development of Alzheimer's disease. *Exp. Gerontol.* **2018**, *107*, 59–66. [CrossRef] [PubMed]
73. Iturria-Medina, Y.; Khan, A.F.; Adewale, Q.; Shirazi, A.H. Blood and brain gene expression trajectories mirror neuropathology and clinical deterioration in neurodegeneration. *Brain* **2020**, *143*, 661–673. [CrossRef] [PubMed]
74. Mizuno, S.; Iijima, R.; Ogishima, S.; Kikuchi, M.; Matsuoka, Y.; Ghosh, S.; Miyamoto, T.; Miyashita, A.; Kuwano, R.; Tanaka, H. AlzPathway: A comprehensive map of signaling pathways of Alzheimer's disease. *BMC Syst. Biol.* **2012**, *6*, 52. [CrossRef]
75. Jia, L.; Piña-Crespo, J.; Li, Y. Restoring Wnt/β-catenin signaling is a promising therapeutic strategy for Alzheimer's disease. *Mol. Brain* **2019**, *12*, 104. [CrossRef]
76. Hsu, W.-C.; Wang, H.-K.; Lee, L.-C.; Fung, H.-C.; Lin, J.-C.; Hsu, H.-P.; Wu, Y.-R.; Ro, L.-S.; Hu, F.-J.; Chang, Y.-T.; et al. Promoter polymorphisms modulating HSPA5 expression may increase susceptibility to Taiwanese Alzheimer's disease. *J. Neural. Transm.* **2008**, *115*, 1537–1543. [CrossRef]
77. Latz, E.; Xiao, T.S.; Stutz, A. Activation and regulation of the inflammasomes. *Nat. Rev. Immunol.* **2013**, *13*, 397–411. [CrossRef]
78. Liu, C.; Cui, G.; Zhu, M.; Kang, X.; Guo, H. Neuroinflammation in Alzheimer's disease: Chemokines produced by astrocytes and chemokine receptors. *Int. J. Clin. Exp. Pathol.* **2014**, *7*, 8342–8355. [PubMed]
79. Zuena, A.R.; Casolini, P.; Lattanzi, R.; Maftei, D. Chemokines in Alzheimer's disease: New insights into prokineticins, chemokine-like proteins. *Front. Pharmacol.* **2019**, *10*, 622. [CrossRef] [PubMed]
80. Moore, Z.; Mobilio, F.; Walker, F.R.; Taylor, J.M.; Crack, P.J. Abrogation of type-I interferon signalling alters the microglial response to Aβ 1–42. *Sci. Rep.* **2020**, *10*, 1–14. [CrossRef] [PubMed]
81. Zhang, H.; Wang, D.; Gong, P.; Lin, A.; Zhang, Y.; Ye, R.D.; Tang, Y. Formyl peptide receptor 2 deficiency improves cognition and attenuates tau hyperphosphorylation and astrogliosis in a mouse model of Alzheimer's disease. *J. Alzheimers Dis.* **2019**, *67*, 169–179. [CrossRef] [PubMed]
82. Wang, Y.; Yang, Y.; Li, C.; Zhang, L. Modulation of neuroinflammation by cysteinyl leukotriene 1 and 2 receptors: Implications for cerebral ischemia and neurodegenerative diseases. *Neurobiol. Aging* **2019**, *87*, 1–10. [CrossRef]
83. Chu, J.; Lauretti, E.; Di Meco, A.; Praticò, D. FLAP pharmacological blockade modulates metabolism of endogenous tau in vivo. *Transl. Psychiatry* **2013**, *3*, e333. [CrossRef]
84. Raychaudhuri, M.; Roy, K.; Das, S.; Mukhopadhyay, D. The N-terminal SH3 domain of Grb2 is required for endosomal localization of AβPP. *J. Alzheimers Dis.* **2012**, *32*, 479–493. [CrossRef]
85. Blair, J.A.; Wang, C.; Hernandez, D.; Siedlak, S.L.; Rodgers, M.S.; Achar, R.K.; Fahmy, L.M.; Torres, S.L.; Petersen, R.B.; Zhu, X.; et al. Individual case analysis of postmortem interval time on brain tissue preservation. *PLoS ONE* **2016**, *11*, e015615.

MDPI
St. Alban-Anlage 66
4052 Basel
Switzerland
Tel. +41 61 683 77 34
Fax +41 61 302 89 18
www.mdpi.com

Genes Editorial Office
E-mail: genes@mdpi.com
www.mdpi.com/journal/genes

www.ingramcontent.com/pod-product-compliance
Lightning Source LLC
LaVergne TN
LVHW070609100526
838202LV00012B/601